Praise for
Grow a New Garden

'Aspirational, accessible, awesome. If only my compost was as rich as this book; Becky's wisdom is palpable. A brilliant balance of the why and the how, this book will pioneer the next generation of gardens. *Grow a New Garden* is the ultimate gardening tool in the shape of a book.'

—**Huw Richards**, author of *The Permaculture Garden*

'Becky's writing is so refreshing. She has learnt by doing and has a refreshing "you can do it" approach. I thoroughly recommend this book to beginners and also to experienced gardeners for inspiring descriptions of simple, beautiful things you can do. Becky demystifies any gardening jobs that look complicated.'

—**Charles Dowding**, author of *No Dig*

'Becky completely captures the passion and excitement of creating your own garden. Her insights from her personal journey combined with expert knowledge truly makes this an accessible book for anyone looking to start, create and maintain their own garden. An ecological approach with nature at its heart is exactly what new gardeners need right now!'

—**Chris Hull**, garden designer and presenter

'An inspiring yet incredibly thorough book covering everything you could possibly need to know about starting a garden from scratch. Becky has the wonderful ability to turn complex topics into easily accessible guidance. This is the book I wish I had had when I first started my gardening journey.'

—**Bex Partridge**, floral artist; founder, Botanical Tales

GROW A NEW GARDEN

GROW A NEW GARDEN

PLAN, DESIGN & TRANSFORM ANY OUTDOOR SPACE

BECKY SEARLE

CHELSEA GREEN PUBLISHING
White River Junction, Vermont
London, UK

First published in 2025 by Chelsea Green Publishing | PO Box 4529 | White River Junction, VT 05001 |
West Wing, Somerset House, Strand | London, WC2R 1LA, UK | www.chelseagreen.com
A Division of Rizzoli International Publications, Inc. | 49 West 27th Street | New York, NY 10001 |
www.rizzoliusa.com

Copyright © 2025 by Rebecca Searle.
All rights reserved.

Unless otherwise noted, all photographs and illustrations copyright © 2025 by Rebecca Searle.
Photographs on pages 19, 35, 52, 54, 60, 83, 113, 114, 119, 124, 142, 147, 148, 153, 176, 178, and 198 (right) by Shutterstock.

No part of this book may be transmitted or reproduced in any form by any means without permission in writing from the publisher.

Publisher: Charles Miers
Deputy Publisher: Matthew Derr
Commissioning Editor: Muna Reyal
Project Manager: Natalie Wallace
Developmental Editor: Sally Morgan
Copy Editor: Angela Boyle
Proofreader: Jacqui Lewis
Indexer: Linda Hallinger
Designer: Melissa Jacobson
Page Layout: Abrah Griggs
Illustrator: Anna Platts

ISBN 978-1-915294-40-1 (hardcover) | ISBN 978-1-915294-41-8 (UK ebook) | ISBN 978-1-64502-365-4 (ebook)
Library of Congress Control Number: 2024055284 (print) | 2024055285 (ebook)

Our Commitment to Green Publishing
Chelsea Green sees publishing as a tool for cultural change and ecological stewardship. We strive to align our book manufacturing practices with our editorial mission and to reduce the impact of our business enterprise in the environment. We print our books using vegetable-based inks whenever possible. This book may cost slightly more because it was printed on paper from responsibly managed forests, and we hope you'll agree that it's worth it. *Grow a New Garden* was printed on paper supplied by Versa that is certified by the Forest Stewardship Council.®

Authorized EU representative for product safety and compliance
Mondadori Libri S.p.A. | www.mondadori.it
via Gian Battista Vico 42 | Milan, Italy 20123

Printed in the United States of America.
10 9 8 7 6 5 4 3 2 1   25 26 27 28 29

This book is dedicated to
my wonderful family, who will always be
the most beautiful things growing
in my garden.

Contents

Introduction		1
1:	What is a 'new' garden?	11
2:	It all starts with soil	31
3:	Plan and design your garden	69
4:	Materials in your garden	107
5:	Creating a nice space	117
6:	Lawn care	139
7:	Basics of plant care	155
8:	Managing pests and weeds	185
9:	Common problems and how to deal with them	205
Epilogue		211
Acknowledgements		*212*
Glossary		*213*
Index		*217*

Introduction

Gardening is often painted as a war. We fight pests, do battle with weeds, build, chop, dig and kill. It is associated with the kind of relentless hard work that recent generations have never had to endure. As a result, we often do the minimum required to ensure that the neighbours don't get upset or to create maintenance-free spaces. A recent trend for garden rooms has blurred the lines between gardening and home improvements. This has been popularised, presumably, due to its emphasis on the finished product and not the continuation of work. The picture-perfect patio is completed with comfy dining set and pizza oven, whilst the use of fake grass or astroturf has risen as people opt to vacuum the lawn.

Our modern world of skyscrapers and Mocha Frappuccinos on commuter trains leaves us with little time to consider the frivolity of enjoying a garden. Millions of us pass by gardens and parks daily without ever stopping to appreciate a nice tree or a larger-than-average bumble bee. Our children, who are naturally inquisitive, are being raised by grown-ups who don't know the difference between a cherry tree and a birch tree. But there is a way to reconnect with nature and allow our children's curiosity to bloom. And it doesn't require as much work as you might think.

Gardening experts are invested in making gardening seem more complicated than it really is. They need to be needed for their knowledge, but the truth is plants want to grow. Most will do what they can to survive, and, simply by using your intuition and caring for them, you will learn how to give them what they need. You will become an expert in the plants in your garden. The trouble is that we have forgotten one really important thing: nature knows what it's doing. Around 3.5 billion

Gardening is about creating balance, not being in a constant battle.

years of evolution have gone into creating plants, their environment and the systems they use to survive and thrive. We started gardening only a few hundred years ago.

I went to a highly religious school in the southwest of England. During the summer of one year, many of my classmates came home from church camp with bracelets sporting the letters 'WWJD' (What Would Jesus Do). These bracelets were allowed, despite the no jewellery rule, because they were deemed to help the children make good decisions. I would like to humbly put in a vote for something similar in gardening and for us to continually ask ourselves: 'What would nature do?'

SMART GARDENING

My generation has been raised to work smart not hard. We were taught to squeeze every available moment out of our day and, when we finish doing that, to see what we could have done better. Shortcuts and 'life hacks' are part of our everyday vernacular. So it's natural that, when we are gardening, we should look for easy solutions to problems. The trouble is that, unlike having AI write your homework for you, gardening requires work and patience.

Organic gardening is often painted as a struggle. Many half-hearted or accidental attempts at organic gardening have bolstered this belief. That time your grandfather forgot to put slug pellets around his cabbages does not count as organic gardening. Furthermore, without the time needed for the practices to start working, it can feel like you are fighting a losing battle. However, the focus of organic gardening is to create an environment where nature does the work so you don't have to. Essentially, it's about letting the natural processes take place to look after the plants in your garden. If you choose to take up the sword and do battle in your

INTRODUCTION

garden, you will be drawn in and the work will be continuous. You cannot hope to fight caterpillars one year and find them all gone the next. For this reason, amongst many other more planet-focused reasons, this book concentrates on organic gardening. I guarantee that gardening with nature is easier than gardening against it.

THE GARDEN ECOSYSTEM

A theme that I will return to multiple times throughout this book is building a healthy ecosystem. So it's important that we first understand what an ecosystem is and why it's relevant to your flower beds.

An ecosystem is a group of organisms interacting with one another in an environment. A healthy ecosystem is an incredible thing: they have numerous systems in place to help them self-regulate. They are fully self-sufficient entities with structures and processes to cope with just about anything. We can think of ecosystems like bodies: they have different parts to complete different tasks. When a toxin enters the body, it is filtered out by the kidneys; and when we fall over, our nervous system kicks in to help us quickly arrange our limbs to minimise impact. These are systems and processes in the body to protect itself. Ecosystems function in a similar way but, instead of having different organs to do the job, they have different species. These can be plants, animals, insects, fungi or microscopic creatures.

Because an ecosystem relies on species and organisms to self-regulate, it is susceptible to damage if they don't have enough diversity. At the same time, however, they are all also much more resilient, adaptable and long-lived than any individual organism. They can deal

Left, Create a garden that thrives with less work and more time for you to enjoy your space. *Right,* All creatures have their part in the ecosystem.

GROW A NEW GARDEN

Your garden is an ecosystem, whether you choose to embrace it or not.

with a wide range of external stresses. Similar to when our bodies have a part removed, ecosystems don't cope so well when species are missing: the ecosystem is incomplete. What has all this got to do with gardening? Our gardens are ecosystems, whether we like it or not.

Unlike rooms in your house, ecosystems are susceptible to forces outside of our control. Some people nurture their garden as an ecosystem, and they will inevitably find more visitors to their garden. This can be both a good thing and a bad thing. Many people don't want to try organic gardening because they feel that it is inviting pests into their garden. However, a well-established garden that has always been cultivated with a view to encouraging wildlife will be a lot less prone to destruction from pests than a new garden that's cultivated in the same way but not yet established. The reason for this? Ecosystems aren't built overnight. They take time. And a healthy ecosystem that has been given time to develop all its systems and processes will be more resilient. This sounds pretty obvious but, in the context of observing your garden under attack from pests, it can be difficult to remember.

Obvious statement number two: the reason that ecosystems take a while to develop is simply that it takes time for creatures to find a habitat. If your garden is in the middle of an estate surrounded by roads, concrete, tarmac and other houses, wildlife will have a harder time finding it than if your house is built in the middle of a forest.

The frustrating thing for us gardeners is that the species we don't want in our gardens – the ones that cause havoc and eat our plants – tend to be the ones that arrive first. These are the species that are specifically adapted to spreading and reproducing quickly. In themselves, they are quite miraculous in their ability to find, inhabit and upset gardens and their custodians. But we're not the only ones that get upset by the presence of these pests; other creatures in the garden need to eat, too. So, if there is one spoilsport species ruining it for everyone else, they aren't going to be too popular. Ecosystems develop defences against these 'selfish' species. These defences usually come in the form of predators or parasites. So, it is perfectly normal and even good to see plenty of pests, at least to start.

Parasites might sound like something you don't want in your garden, but the right parasites are vital pest control – they just need a better PR team. The thing with predators and parasites is that they will not necessarily be present in an ecosystem that does not have the prey or host species present in it.

Predators and parasites are not likely to show up to your garden and wait patiently until you have some caterpillars for them to eat or lay eggs in. Therefore, we must first go through the pain of enduring these pest species (to some extent at least). This period will help our ecosystem get established and create less

Top, Mature gardens have mature, well-established plants. *Bottom,* In an urban environment, it will take longer for species to find your garden and begin establishing an ecosystem.

GROW A NEW GARDEN

Left, To attract predators such as ladybirds, there must first be something, such as aphids, for them to eat. *Right*, This tiny ecosystem is perched on top of an old fence post.

work for us when we are fighting on the front line of pest defence by employing some much more capable individuals.

There is a catch, of course: gardens are generally quite small and can't function as proper ecosystems all of the time. On a small scale, they can. Ecosystems can be as tiny as a patch of lichen, a pot of soil or a tree, after all. But the garden, as an isolated environment, will always be more vulnerable than a sweeping forest.

A LITTLE ABOUT ME

Gardening has been in my blood for as long as I can remember. I used to love the freedom of lying on my stomach, examining tiny creatures as they went about their lives. I would mentally plan how my garden would look when I was older (there would be a lot of roses, a good tree for climbing, and the borders would go back as far as the eye could see). As soon as I got my first garden, I was coiled and ready to spring into action.

INTRODUCTION

Being naturally curious about the world around me and being raised by a mother who was similarly curious, I fell into studying ecology. My mother changed careers to ecology when I was in my early teens. Suddenly, developing a career in lying on my stomach examining tiny creatures felt entirely possible. Through my degree, I noticed myself questioning things around me more and more. Discovering that everything had a purpose, and even if we didn't know what it was, there must be an answer to the question of why, wherever it may arise.

This questioning followed me to the threshold of my first garden. A part of me wanted to disconnect my brain entirely, and go into auto-pilot, dutifully following the second-hand gardening advice I had picked up through my life. But the analytical side of me started to ask why at every turn. It developed into a temporary paralysis when I realised that, in the Venn diagram of gardening and ecology, the overlap was minute. So I decided to garden my own way. Of course, thousands if not millions of people worldwide grow plants and garden like me: in a way that seeks to be positive for the environment and for the gardener. It's less work, and, instead of tackling problems directly, we ask why they are occurring in the first place.

Since this first garden, a series of events has led me to starting numerous new gardens. I have moved around so much in my adult life that I inadvertently gained a lot of hands-on experience. This practice came to something of a head when I finally got my hands on a nice, big, mature but overgrown garden. It was around a quarter of an acre, with an old pond, a shed, a greenhouse and multiple mature fruit trees. It was through this garden that I found my way into the world of horticulture and became a garden writer, speaker and social media creator. Nowadays, my full-time job is demystifying the world of gardening to make it more accessible for others.

Gardening, to me, is the ultimate way to feed my curiosity.

GROW A NEW GARDEN

Throughout my work, I always ask *why*. The gardening advice that I give takes into consideration natural processes and seeks to harness them, making our gardening easier and more sustainable.

A NOTE ON ALLOTMENTS

I will mention my allotment from time to time. Many of the gardening principles in this book can be applied to creating an allotment as well. Allotments are as much a part of the English landscape as country cottages and village pubs. They are small plots of land that are usually publicly owned. They are divided up and rented out to anyone wishing to grow their own food for home consumption (not for commercial production). Allotments are generally quite inexpensive and a great way to make up for the fact that our houses generally don't come with much land. They are essentially community gardens, but everyone has their own slice of it to do with as they please. Mine is around 100 square metres, which is small for an allotment but plenty of space for me to use to grow food and flowers for my family. Allotments are generally quite sought-after, and the waiting lists can be many years long. A lot of people have an allotment for their entire lives, if they are lucky enough to secure one.

If you are on an allotment waiting list, there is a way to push to get a new site opened. Legally, the council has to provide you with an allotment if there are six or more of you who do not have access to an allotment and want one. This can be like getting blood out of a stone, but I know from experience that with a little persistence it is possible. Our own allotment site was due to be put in by the developer and was eight years overdue (it was very low on their list of priorities). When my partner and I moved to the area and realised that there was an allotment site right across the road from us that was allocated but not being used, we started writing letters. We wrote to the council, the local councillors and the developers. We gathered almost a hundred signatories and put some real pressure on them. It took a further year, but the allotments were finally 'opened, and now 15 new plot holders are happily growing' their own vegetables.

This is my allotment, a place where I can grow food and not worry about what it looks like!

CHAPTER 1

What is a 'new' garden?

For the purposes of this book, a *new* garden is any patch of land that is unknown to you, that you intend for use as a garden. It might be that you have just moved house, so the garden is entirely new to you. It may be that you are about to overhaul an existing space. If the garden you are about to create was paved or decked over or covered in weeds or (the dreaded) astroturf, you don't know what you will find. Whatever the circumstances, though, I congratulate you on your new and exciting project.

There you stand on the threshold of greatness. Wearing slippers and dressing gown, hot cup of tea in hand, today is the day you start your new garden. Your hand instinctively goes to your back, knowing that, when you walk through that door later, it will be aching. If you're like me, you will be drinking that tea a little slower than usual. It's not because you don't want to start but because you know that, once you do, you're on the journey and there is no turning back. Your mind is running at a hundred miles an hour, trying to decide what to do first, if your new dahlia will be happy in that corner, whether you can even afford to take this project on right now. Standing there, you see something that is less than inspiring but more than exciting.

Even if you have been in the house and around a garden for many years, and life has got in the way of sorting it, your old garden could still be a new garden if you decide to have an overhaul. What is more likely is that you have recently moved house or taken on this outside space, and you are hoping I might be able to tell you where to start.

The new garden I took on in 2018.

GROW A NEW GARDEN

The best thing about starting a new garden is that the garden has a will of its own. It will pull you in different directions, and you can rarely second-guess the thrills and challenges you will face. I have had the fortune – or perhaps misfortune – of starting numerous new gardens. Each one has brought its own unique set of joys and troubles.

The day I first stepped out into my little terraced garden in Southwest England, I had a head full of dreams. None of them particularly coherent. My problem was not lack of inspiration but rather lack of funds, and I was living in a rental property. My landlords were OK with me doing anything, providing it was restored to grass afterwards. So, I started with my primary concern: I wanted to grow food to feed my children. Regardless of how it looked, it must produce food.

With the help of my dad and a flappy, plastic greenhouse, I filled the garden with vegetables. My budget was zero so I

Top, The garden was put together using materials I got as cheaply as possible and designed to be removable afterwards. *Bottom,* After three months, the garden was brimming with organic veg and the pests hadn't even found my garden yet.

WHAT IS A 'NEW' GARDEN?

started by scrounging some decking boards that had been over-ordered by a neighbour of mine. I got a job at a local farm, helping out with the lambing, and collected all the horse manure my car would carry – much to my children's disgust! And thus, out of poop and timber, my new garden was born. In truth, I didn't need the timber. I opted for raised beds for the simple reason that I have a naughty little tortoise, who enjoys eating my plants. As it turns out, he is also an accomplished climber, and this did little to keep him out!

TYPES OF NEW GARDEN

There are several different types of new garden and we will look at a few here. Don't feel you need to pigeon-hole your garden into one of these categories, but having a clear idea of what kind of garden you're dealing with in the first place can help you know where to start. It might also help you feel a little less alone in your venture.

New-build gardens

There are, of course, some unique challenges that we can face when starting a new garden with a newly built house because these kinds of gardens are rarely given much regard by the developers. They often look good when you first move into the property but, as time goes on, problems can arise. This isn't the case with all new-build gardens, but on the whole, if you are starting with a new-build garden, you will be on the back foot.

This challenge is primarily due to the way that houses are built, especially when there is a large group of houses all being built together. The developers start by removing the topsoil. This step is so they can bring heavy machinery onto the site. It makes good sense in the context of building houses and not losing your cement mixer to a sticky gloopy mess, but in the context of a garden it doesn't. Topsoil is the part of the soil that matters most to us gardeners. It's where the nutrients are, the water is stored, the plants' roots will live, and it's part of the soil that is easily dug. It's also the part of the soil that stops water from gathering on the surface, because it absorbs water.

Below the topsoil, we have the subsoil, which is much more solid. It's the part of the soil that, when your spade reaches it, makes you question how much you really want to do this. It can often feel like

GROW A NEW GARDEN

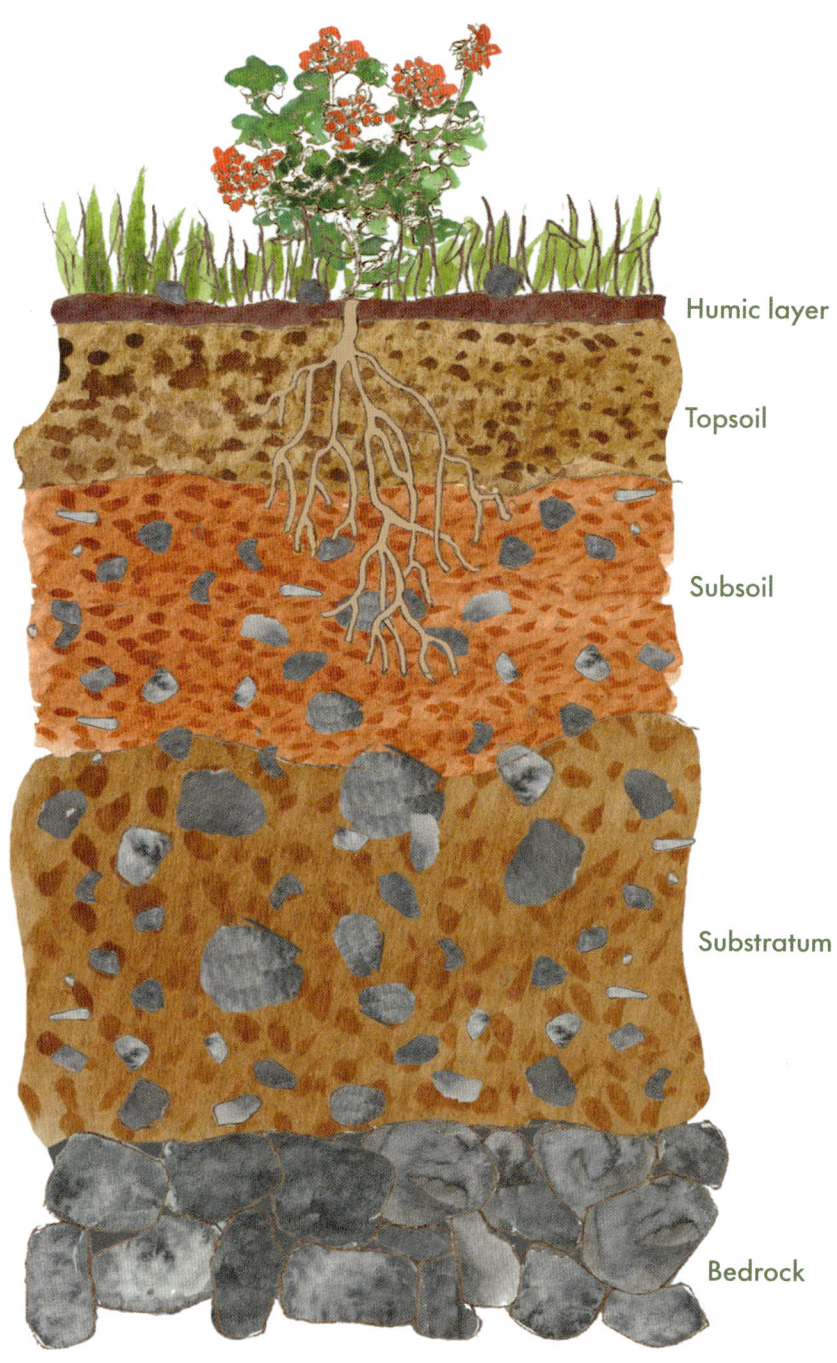

Layers within the soil. On top of the topsoil is the 'humic layer', consisting of dead organic matter.

A new-build garden.

you've hit something completely solid. Whilst not great for gardeners, it is superb if you happen to be driving a 30-tonne crane. When the houses have been built, the original topsoil is often shipped back onto the gardens. But if the developer is trying to cut corners or if the current soil 'doesn't look too bad' to the site manager, they won't bother. However, a house cannot be sold with a bare patch of mud outside the back door. So, the developer will get some turf and lay it on the garden. This rarely involves even putting a spade into the soil to check whether it's OK. From time to time, the unlucky amongst us will even find builders' rubble underneath our gardens; it can be quite costly to get rid of rubble, so why not simply bury it? Again, this makes total sense in the context of building a house but no sense at all in the context of making a garden.

The result of this developer mentality is that we can have extraordinarily hard and often polluted soil. We are essentially dealing with subsoil rather than topsoil. Even with topsoil, the effect of building can be very negative. It can lead to problems with flooding or drying out. In some cases, it can mean just digging a hole for a plant is an excruciating job. It can also mean that plants put into the ground fail to thrive. To learn more about this, see 'It all starts with soil' (page 31).

Overgrown gardens

My favourite type of new garden, the overgrown garden, can be a treasure trove of discovery and delight. It looks intimidating, swathed in weeds and thorns. You might not even be able to tell how big the garden is right now.

GROW A NEW GARDEN

When I was a student, my friends and I moved into a small semi-detached house with a decent-sized garden. We dreamed of fun summer nights sitting in the garden, drinking beer and kicking a ball around or sitting by the fire telling stories and jokes. The garden – being that of a student house – was not the sort that you might see on television, with beautiful people draping themselves over expensive furniture. It was overgrown, with a very small patch of grass. No problem, because we were young, fit students. We gathered up what gardening tools we could find – a rusty pair of secateurs (pruning shears), a rustier pair of loppers and a spade – and off we went. Roughly two metres from the back door was an impenetrable bramble fortress, which we gleefully hacked through, collecting the branches up as we went. Brambles aren't the friendliest of plants, with their large, strong thorns that easily pierced our budget-friendly gardening gloves. One by one, my housemates dropped out of the working party until it was just me and one friend. During the course of a few hours, he and I fought our way through several metres of brambles, trimmed back the hedges, discovered some nice trees and tripled the size of our garden. That evening we all sat out and enjoyed a few beers around a fire of dry, old sticks we had found during the course of the day.

As you can imagine, after a single day of bramble bashing (which, knowing us, probably didn't start until well into the afternoon), it wasn't exactly Kew Gardens, but it was all we needed, and it didn't cost us a penny. Once the weeds were cleared, we were able to properly observe the garden. In spring, a few brave daffodils pushed through the ground and the grass started to grow back. What we were left with was not an overgrown garden but a neglected garden.

The best thing about starting with an overgrown garden is that incredible sense of achievement you feel when you have cleared it. Also, you get plenty of compost material to start you off in your new space. I highly recommend protective clothing if you are dealing with unfriendly spiking, stinging or irritating weeds. You and your garden should start off on a positive note, rather than with you cussing it for injuring you!

The hardest thing about an overgrown garden is that you will struggle to start planning until after you have cleared it at least some of

An overgrown garden.

A neglected garden.

the way. You need to be able to see what you're dealing with before you can make any solid plans.

Neglected gardens

These are the gardens that have some plants, some grass, but not much else. Nothing seems particularly well thought out, and nothing has been well maintained. You can see where your boundaries are and how big the space is. There will be a few surprises in neglected gardens; plants that pop up unexpectedly, buried 'treasure' in the form of random things left in the garden through the years, or discovering hidden delights, such as a pond, that have been slowly obscured from view by years of neglect.

These gardens come in a few shapes and sizes. Some are quite overgrown, but the distinction is that you can access all or at least most parts of the garden. This means that you can start planning right away. It might be slightly more of a challenge to plan than a blank-canvas garden. This is because you will want to identify any plants that you want to keep or any that you want to remove. With the exception of trees and climbing plants that don't take kindly to being moved, you can to some extent employ pots for this job of rearranging. Pots or

WHAT IS A 'NEW' GARDEN?

containers can act as temporary holding pens for plants whilst you decide if you want them or not. You can then put plants back, move them around or pass them on to someone who does want them.

Neglected gardens may also be completely devoid of life, save for a few brave weeds pushing through the cracks. They can be completely paved over, or covered in deck, or (heaven forbid) just concrete. Either way, the chances are, if you have a neglected garden, you have your work cut out for you. The chances of discovering something unexpected are also quite high. But the impact you can make is also much greater; just imagine the before and after pictures.

Not-your-style gardens

You have a garden, and someone has made some effort with it, but it's really not your thing. You have three choices: you can learn to live with it and add your own unique touches to make it feel like home; you can jig it around a bit and try to use what you can to make your new garden; or you can rip it all out and start again. These three options are for very different budgets. And there are positives and negatives to each. Ripping it all out and starting again will inevitably cost you a lot of money, even if you do it yourself, and there is a chance you could find some nasty surprises. However, you will get the garden you want, and, if your budget is healthy, why not go for it? Consider if there are any elements you can repurpose and personalise. Or if you really don't want to reuse, is there any of it you can sell to fund the next stage of your project? It's a good idea, if the budget is tight, to consider a staged approach to revamping your garden. Putting some cheerful containers in a corner is a good way to press pause on an area without it upsetting you to look at it.

A garden can be beautiful, but if it's not your style, you may need to start over.

Whatever the garden you are taking on, you will encounter some unexpected challenges and some unexpected joys. Remember that starting a garden can be done relatively quickly, but truly creating or curating a garden can take years. It's a journey through time, weather, space and nature. It can take you in many different directions over the years or could be something you do in a weekend and only return to with a margarita in hand and a good book.

Not-a-garden gardens

If you are starting out on a patch of earth that has never before been seen as a garden – perhaps it has just been reclaimed – you have this type of garden. This type will come with its fair share of surprises, no doubt, but it can be quite a gift. Particularly if the area was widely untouched before you got there, it's likely that weeds will be your first and probably main adversary. The superb thing about these types of gardens is that you can make a huge difference to them. Plus your before and after pictures will be guaranteed to go Instagram viral (if that's what you're into).

STRUGGLES OF A NEW GARDEN

Each garden is unique. Whether you are stepping out into a garden for the first time or making big changes to a garden you are well-acquainted with, your new garden will come with some challenges. If we learn to anticipate these challenges and accept that they are an inevitable part of building a new garden, we can be proactive.

Whether you are yet to become a gardener or you have decades of experience, every new garden will come with some challenges. Each garden is unique, and there are a lot of ways in which they can differ from one another. For example, my garden has two levels, and both are flat, whilst the garden next door slopes downwards. My garden has drainage problems and regularly has standing water on the lawn. My auntie, who lives just down the road, has totally different soil and constantly needs to water her plants to stop them from wilting. My new garden has a friendly squirrel that eats vegetables and digs holes in the lawn, but my last garden had more of a problem with the neighbourhood cats. There can be a tremendous amount of variation even within a garden, with some areas shaded, some in full sun, some cool,

WHAT IS A 'NEW' GARDEN?

some naturally warm, others wet and some dry, and so on. In short, designing a garden is nothing like designing a room in your house. And you have to be willing to accept that some things just won't work in your space. The flip side of this is that it's much more exciting, plus you get the endless joy of sharing your space with nature.

Pests

When we start a garden – or do something different in an existing garden – the neighbourhood wildlife will be delighted. There is something new on the menu and they want to try. We can't blame them for this. They're hungry, and we're growing their food.

Here's the thing: your garden is an ecosystem whether you want it to be or not. Some gardens are pretty desolate, with about as much biodiversity as a car park, and others are thriving jungles of activity. It's a spectrum. Since many people struggle with pests when they first start a garden, it's natural to want to get rid of the pests. Many people do that by trying to eliminate as much life from their gardens as possible. It does feel from time to time that everything that crawls, flies, wriggles, walks or slithers through the garden is against you, and so the battle that is gardening begins. In reality, what is happening is that all the creatures, from tiny bacteria to the larger animals in your garden, are rejoicing! You have started a garden and finally there is something to eat. You will feel this much more if your house is surrounded by other houses with barren gardens; paving stones, gravel and fake grass will all serve to chase the wildlife out of their gardens and into yours. The problem is that inhabitants that you don't necessarily want in your garden are usually the most resilient and the ones most determined to stay.

Pests can be a big problem in new gardens.

GROW A NEW GARDEN

Imagine you're a tiny insect, like this cucumber beetle, looking for food.

Imagine for a moment that you are a tiny insect. Life is pretty good in the little field you live in. There is plenty of food and plenty of water so, like any self-respecting insect, you reproduce. Now imagine a building developer comes along and puts houses on the field where you and your insect spouse are raising your children. Just when you think all is lost, the developers put in some gardens, small patches of grass. Some of the gardens are paved, which is not good for you at all, so you have to stick to the ones that have grass. Of course, there are so many of you by now that you might be causing problems on the patches of grass. You don't mean to, but that's just the way it goes. After a while your population starts to go down. Then a human moves into one of the houses and decides to start a little vegetable patch. She grows some flowers, plants a tree and makes a little pond. If you're extra lucky, she might even make a log pile and a compost heap. This is probably the best thing that's ever happened to you, if you're an insect. It's no good to you if you live three gardens away; you might be too small to even get there. But if you are big enough or mobile enough, you will most certainly take advantage of this brand-new oasis.

Now, let's venture a little closer to home and 'imagine' that we are the gardener who just set up the garden. Despair. All your hard work to create something beautiful and wildlife friendly, and how do they repay you? By eating everything! It's just not fair. If it's your first garden, you might consider that, perhaps, you've done everything wrong or maybe that your garden is just in a terrible location for a

WHAT IS A 'NEW' GARDEN?

garden, and that it's overrun by pests. Worse still, you might reach the conclusion that gardening is really hard and, actually, you didn't want to do it anyway.

You are left with two choices. You can give up and admit to your smug granny that their generation is just better at these sorts of things. Or you can find out what went wrong. If you've picked up this book before starting a new garden, please don't let me put you off. What I'm talking about is simply nature balancing itself out. It's a process that has to happen, but there are some ways that we can mitigate against the heartache that can ensue. So, the first struggle of a new gardener is always going to be discovering the various problems that are specific to your garden. Whether this be pests (which, by the way, might not happen), flooding, drying out, lack of sunshine, too much sunshine, next door's cat using your garden as a toilet or the various other things that can go wrong. One of the most important things is to be prepared for this. It's absolutely acceptable and to be expected, that you will encounter some difficulties. If it's pests, put yourself back into the mindset of the little insect with his little insect spouse, and ask yourself, 'What would I do?'

There are several ways that we can mitigate the various problems we might encounter in our gardens. The first is to accept that it's probably not your fault. The second is to accept that there is probably something you can do about it. The third is to remember that, even if you are the most experienced gardener, things will still go wrong. The main thing standing between your garden as it is now and your garden looking amazing is time. You will notice that none of these comments are actually gardening advice but rather something that you can do right now whilst you read this book. And something you will need to continue to do as you progress through your gardening journey.

Fortunately, there are also so many practical ways to get the garden that you are dreaming of whilst enjoying yourself and not letting it become an all-you-can-eat-buffet for insects. The rest of this book will endeavour to cover some of the things – both joyous and not so joyous – that you might face in your new garden.

Mindset

One thing that many people can struggle with when they're starting a new garden is becoming overwhelmed. I'd be prepared to bet that this one bump in the road stops a lot of people from even trying. Let's be honest; it's no mean feat creating a garden. But it doesn't need to be back-breaking, soul-crushing work. By taking your garden design or redesign at your own pace and creating realistic goals about what you can achieve, you can help to combat the feeling of being overwhelmed and relax into the process. I often find that, with larger or completely bare or overgrown spaces, it is easier to divide the space, either physically or mentally, before you begin.

It is worth having a clear idea of what you would like to achieve in your garden before you start. You can try making a sketch of it, or just creating a mental image of something that you would like. Scrolling Pinterest and Instagram can help you create a mood board for your garden, and this will in turn give you something positive to focus on. I'm going to attempt to take you through the easiest path of creating a garden. Sometimes there will be shortcuts; sometimes a shortcut does not exist. Other times, the shortcut is more expensive. But one thing is for sure: if you are not enjoying the process, and you do not have a clear vision for what you would like in the end, go back to the beginning and ask yourself why you are doing this in the first place. If the answer is 'I have no idea' whilst you are sprawled exhausted on the sofa, with mud under your fingers and an aching spine, it's probably just time to take a break. But remember, anything that is worth having is worth working for.

If you have a plan of what you want the garden to look like, you can also create a plan for how you will tackle each part of the garden. Set yourself realistic goals, and don't take on too much at once. Getting one small part of your garden up and running (and hopefully filled with beautiful plants) will give you the boost you need to get out there and do more. It is also a good opportunity to learn from your garden and adjust your expectations accordingly.

IMPORTANCE OF MAKING A GARDEN

A garden is good for your health, both mental and physical; it's good for wildlife (when done with wildlife in mind); it can bring joy to those around you; and it can even help fight climate change. A garden can bring you free food and flowers; it can be a home for animals; and it

My garden is my place to escape from the world.

can be somewhere for you to connect with the world around you. A garden can be used to entertain when you run out of space in your house, or it can be used to soak up some sun in the summer. It will be the place where your children first hold a worm or see a butterfly. It is a place you can go to calm down or to cheer up. The versatility of a garden is almost unending.

You are already incredibly fortunate to have a garden. Whether you own it, rent it or are starting at a community garden, it is a privilege. And you have a duty of care to that place. It might be the only outside place that you get to take care of, and what you do with it does really count.

Health

Most of the things our doctors and other health professionals tell us to do to maintain good mental and physical health can be found in a garden. Gardening is active, mindful, outdoors, grounding, and can provide us with healthy, nutritious food. Mental health has even been known to improve in individuals who just *see* nice gardens. So, you could even be improving the mental health of others in your neighbourhood, simply by creating a garden.

Biodiversity

When it comes to biodiversity, we can campaign, donate to charities, share articles, take part in community projects and teach our children about its importance. But for most of us, the only place that we have full control over enhancing biodiversity is our garden. So, if you want

WHAT IS A 'NEW' GARDEN?

to do something really positive, the best place to start is your garden. You might look at your garden, particularly if it is urban or surrounded by houses, and wonder how it could possibly make a difference to biodiversity. But in these cases, your garden is possibly even more important. It is an oasis in the desert for species passing through, and a haven for those who have been otherwise displaced. No matter how small, or uninteresting, your garden can be improved for wildlife, and can be an important refuge for nature.

There was a beautiful study conducted by Jennifer Owen in her small suburban garden in Leicester. Jennifer counted and recorded all the species she found in her garden over the space of 30 years. Remarkably, she counted 2,673 species, from plants to insects and mammals. This is an astonishing amount of biodiversity for a small space and, indeed, much more than one would expect to find in most natural environments. The direct contribution that gardens make to biodiversity is significant, and you will notice it in your own garden once you start.

My humble garden is surrounded by houses and has only one entry point for creatures, such as hedgehogs, that can't leap, climb or fly. When we moved in, it had three mature birch trees lining the back fence and not much else. We soon noticed the birds enjoying the catkins on the birch trees, and in February of our first year we put up a bird box on the back fence. Within 24 hours there was a pair of bluetits scoping it out. They ended up only using it like some sort of seedy motel to do their business and then leave, but the following year they took up residence and raised their young.

Left, Producing fresh food is just one way gardening improves our health. *Right*, Gardens have enormous potential to contribute to national and global biodiversity.

GROW A NEW GARDEN

By then we had put up another bird box, several bird feeders, a few bee hotels and a butterfly hotel, made a dead hedge and filled the garden with plants. There was a compost heap, two ponds and a custom-made hedgehog house. During our first year in the garden, and since putting these features in, the surge in wildlife has been noticeable in the entire neighbourhood. In fact one of my neighbours, who, unbeknown to me, was following me on Instagram, actually messaged me to say that she has seen so much more wildlife in the neighbourhood since we moved in. This increase isn't exactly tested and confirmed, but there is certainly a cheeky squirrel that visits now, a family of hedgehogs, a host of birds that feed from seed heads in winter and bathe in and drink from the ponds. There is a fat pigeon that comes every hour, on the hour to take a few big gulps from the pond and then flies off to the nearby oak tree again. Before we made a pond in our garden, I'm not sure where the nearest reliable source of water would have been for these animals. Without knowing if my neighbours have ponds (I suspect most do not), I can't give an accurate estimate on this, except there is a little stream about half a mile away. If you're a small animal without any means of storing water, you will certainly not want to be too far away from somewhere you can drink. It's not like birds can carry around little water bottles. So simply by creating a pond, we opened up more space for these creatures to explore and live. It's like adding another service station to a long and desolate highway.

Gardens are also a fabulous opportunity for us to connect with the nature around us, and for our children to connect, too. Those of us who grew up surrounded by birds and butterflies will instinctively care more about these things. We will notice their absence more profoundly, and we will call them by their names. By learning the names of some of the birds, butterflies, beetles and mammals that visit your garden, you can begin to learn their unique characters and how important each species is. By creating a garden for ourselves and our children, we create the opportunity to learn about nature through observation. If you're over the age of 30, or just are inclined to enjoy these sorts of things, there can be a pure joy gained from watching birds in your garden. It takes you by surprise sometimes. One moment,

My humble pond, after just one year, is proud to be home to a handful of newts, some dragonfly larvae and a lot of pond snails, and is frequently visited by other creatures.

GROW A NEW GARDEN

Joy should always be a central part of a garden, however you choose to approach it.

you're young and cool, and the next you're delightedly telling your partner that you have a goldfinch in your garden.

Joy

Joy is, perhaps, the most important reason to create a garden. Yes, a garden is good for your mental health. It'll probably increase your chances of being able to sell your house. It might get you fit or give you fresh fruit and vegetables, and hopefully it will bolster biodiversity. But I can guarantee that your garden will bring you joy. If you haven't been through the process of creating a new garden before, it's hard to describe the joy that it brings. Gardening is certainly a process and not a destination, but when you first sit down to enjoy the flowers you will feel like you have arrived. All your hard work will be forgotten when you first spot a bee landing on those flowers that you planted. When a bird takes up residence in a newly hung nest box, it will be exciting: they're here because you're here. Inviting your friends over for a barbecue and sitting in the garden on a warm summer's evening with a bottle of wine, telling them that you could've sworn that the dahlia was going to be yellow but you're happy that it turned out orange. Watching your children running around and coming inside later with little, muddy fingers, or eating sun-sweet fruits. However you choose to enjoy your garden, if you create it yourself it will be a space that is full of joy.

CHAPTER 2
It all starts with soil

If you're not a gardener yet, and you have never unashamedly added horse manure to your Christmas list, you might be tempted to skip this chapter. But even if you skip the rest of the book, I implore you to read at least this chapter. This one is by far the most important in the whole book and contains information that most gardening books don't. If you have come here specifically for this chapter, then I welcome you, and I hope it doesn't disappoint!

I am what my family would call (if they were honest rather than kind) an 'insufferable bore' when it comes to soil. I genuinely believe

Left, It all starts with soil. *Right,* Plant roots have an intimate relationship with the soil.

Healthy soil means healthy plants.

that it is one of the most important things on Earth. I don't think it would be right to single out one thing because, as an ecologist, the first thing I learned is that everything is connected. We can't have soil without trees, and we can't have trees without soil.

Soil, however, is responsible for an awful lot more than we usually give it credit for. Gardeners tend to understand that soil holds water and nutrients that our plants need, and helpfully provides our plants with anchorage so they don't topple over. But soil is also the largest freshwater reservoir on the planet, the biggest carbon sink, and the most biodiverse ecosystem in the world. Soil is responsible for producing 95 per cent of the world's crops, and almost all terrestrial life relies on the top 6 inches of soil.

In our gardens, healthy soil can:

- regulate water levels to help prevent flooding and drying out
- feed our plants without the need for fertiliser

IT ALL STARTS WITH SOIL

- help combat pests and diseases in and out of the soil
- bolster biodiversity and help foster balance in your garden ecosystem
- manage its own pH
- reduce weeds

This all sounds quite exciting, but it's difficult to understand how this works and, therefore, difficult to trust that it works unless you first know a little bit about soil.

SOIL IS AN ECOSYSTEM

We gardeners obsess about soil. We talk about it, we read about it, we complain about it, and we work it. We add to our soils, we dig into our soils, we test our soils, and we plant in our soils. But very few of us actually understand soil (though soil is becoming more of a hot topic these days, so more are starting to understand it).

The first thing that you must know about soil is that it is a living, breathing ecosystem. It's a collection of plants and animals interacting with one another in their environment. In fact, there are more tiny organisms in one handful of healthy soil than humans who have ever lived on the planet. I give talks about this all the time, and the one thing I hear over and over is, 'Not my soil. Mine's totally dead.' Well, let me reassure you: unless your soil is submerged in water on a constant basis, has been put through a microwave or has been drenched in chemicals, it's probably not totally dead. Most soils can be reinvigorated and rejuvenated.

The soil ecosystem functions in much the same way as any other ecosystem. The only difference is that most of it is microscopic. It is for this one simple reason that it has not been observed as an ecosystem until relatively recently. Incidentally, this is also why some people believe their soil to be completely dead. Just as in every other ecosystem, the soil contains a dazzling diversity of creatures, all occupying their own specific niche. In doing so, many of them are also maintaining the environment around them.

As children, we are taught about food chains. I had the distinct honour of teaching my eldest about food chains during lockdown. We are taught things like a caterpillar eats plants and a bird eats a

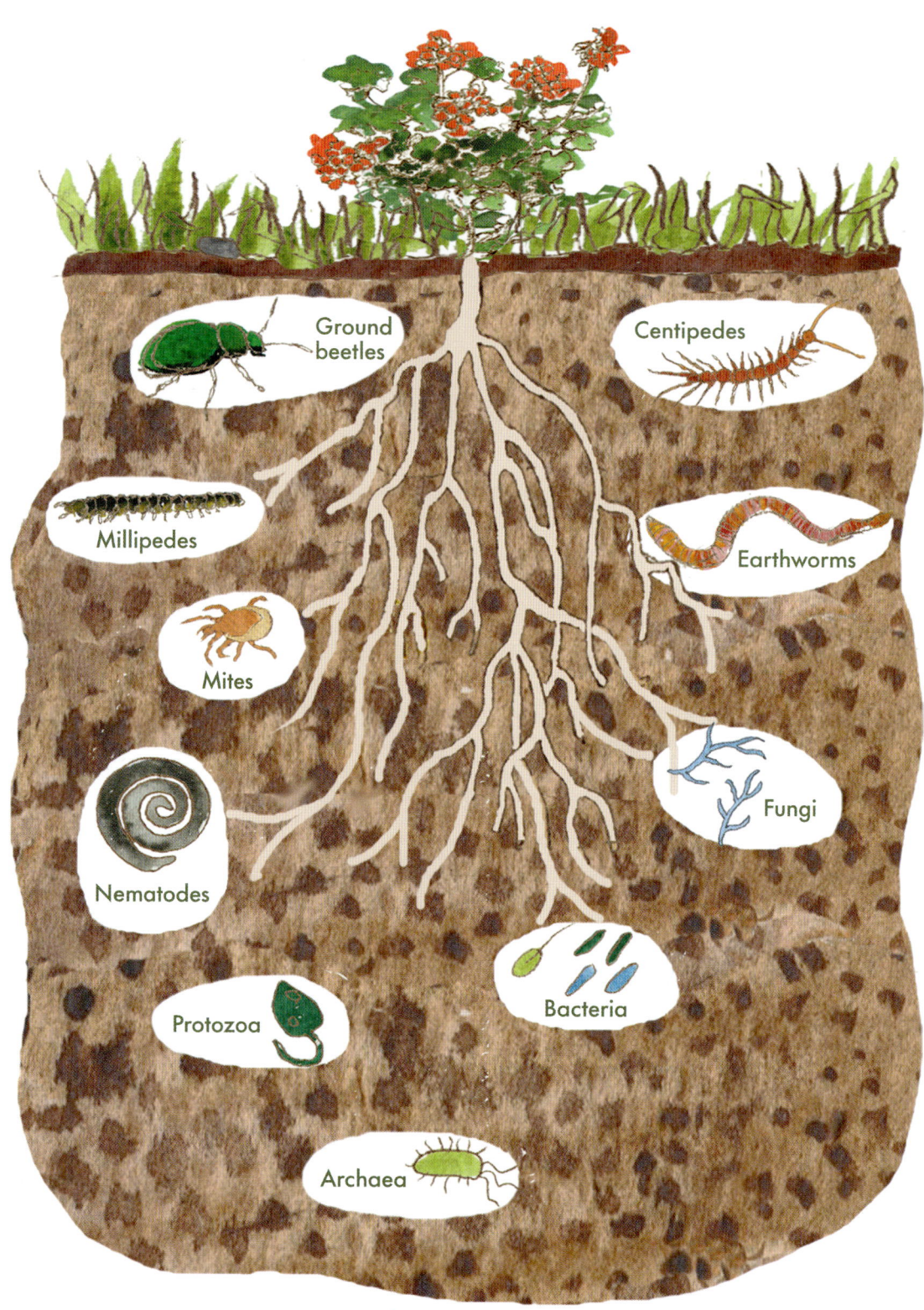

IT ALL STARTS WITH SOIL

caterpillar. Of course, we gardeners know that there are many more things that eat our plants than just caterpillars, so things are never as simple as they are in a food chain. This description is merely a representation for young children to understand the idea. Closer to the truth is what is known as a food web. These diagrams demonstrate the interconnectedness of multiple animals and plants. But what food webs and food chains have in common is that they both start with plants. The reason for this starting point is that plants are the only primary producers. This means that they turn thin air into food or gas into mass. Another way of looking at this is that there is no food on the planet that doesn't, at some point, come from plants. That's right, even Mars bars and HARIBO Goldbears, which I consider to be very far from plants, at one point along their lifetimes were just that.

There are so many creatures that live in the soil – earthworms are really just the poster boys.

Inside the soil exists a delicate and complex food web, and, like other food webs and food chains, it starts with plants. The life in the soil feeds primarily on dead leaves. But there is another plant-based food source for the soil food web, and it comes in the form of something called *root exudates*. These are little packages of sugar and other carbohydrates that plants produce during photosynthesis and pump down through their roots into the soil to feed the organisms in the soil. This exchange shows us something important about the soil ecosystem. The organisms in the soil are not taking it from the plants; the plants are giving it away willingly. In fact, some plants give as much as 40 per cent of the carbohydrates they produce. This gift is pretty baffling until you understand why they do this. You see, the most important thing to understand about soil is that plants and the soil

The soil food web: nutrients are cycled through the soil by the life in the soil, which in turn creates good soil structure.

ecosystem exist in symbiosis. This means that they have a mutualistic relationship, where both the plants and the soil benefit. So what are the plants getting out of it?

The soil ecosystem actually does an awful lot for plants. Firstly, it helps to maintain soil structure. Good soil structure can hold on to water without getting waterlogged and hold on to air so that plant roots can respire. The life in the soil also helps to release nutrients to our plants and offers protection from a few organisms that live in the soil and can cause them damage, such as pests and disease-causing microbes.

HOW PLANTS GET THEIR NUTRIENTS

Most soils contain plenty of nutrients, but they are locked up in the organic matter, sediments, particles and even the bedrock of the soil. Unfortunately for plants, they do not possess the necessary tools to extract these nutrients. Plant roots can only draw up water and nutrients from the soil like a straw. When my children were little, I would give them little smoothies in cartons. They loved the smoothies but weren't able to get into them. It was up to me to take the straw and pierce the little foil seal so that they could drink. I was the soil organisms unpacking the nutrients so my plant children could suck it up through their straws. If nutrients in the soil existed in a form that could easily be dissolved into water and taken up by plant roots, a few decent rainfall events would see all the nutrients in the soil washed away. Therefore, nature has devised a strategy to keep the nutrients in the soil and allow plants to have access to them. And that is where the soil food web comes in. They're the ones that pierce the foil for the plants.

Organisms in the soil can break down organic compounds and chemical bonds, allowing the nutrients to be dissolved into water and thus releasing them to the plants. The mechanism used for breakdown differs between different nutrients and different organisms, but the basic principle is that nutrients are released by everyday eating and excreting carried out by organisms in the soil. Each one plays a slightly different role. It's important not to get bogged down in the details of this – exactly how it works really doesn't matter to us as gardeners – but it is incredibly important to

IT ALL STARTS WITH SOIL

understand that plants cannot get nutrients from the soil without the microorganisms in the soil.

For many years the question of why the soil food web would offer this service to plants utterly baffled scientists. But now that we understand about root exudates, it makes complete sense. The organisms in the soil want to take care of our plants, want them to be healthy, so they can get their delicious root exudates. This partnership goes so far that plants regularly use chemical signals sent down through their roots to request specific nutrients when they need it. This is an oversimplification of how the process works, but essentially a plant will produce and use special sugars and other carbohydrates to attract organisms that can help them get hold of the nutrients they need.

This request process means that plants can have the nutrients they need before they start to show signs of deficiencies. Standard modern-day gardening advice would be to feed your plants with liquid nutrients. We can only discern if this is not working correctly or if a plant is missing out on a particular nutrient by detecting discolouration or changes in growth. Not only does this require us to be extraordinarily aware of what our plants are doing, but it also requires us to know how to react. If our plants get nutrients through the natural processes in the soil, there is no need to second-guess their needs. So, we can instead put our efforts into building a healthy soil ecosystem, which is a far simpler task!

Plants work with the life in the soil, including fungi, to get nutrients in exchange for the sugars and other carbohydrates they provide.

GROW A NEW GARDEN

The other problem with liquid plant feeds is that they interrupt the reciprocity between the soil and the plant. If a plant can get its nutrients without having to feed organisms in the soil, it will. So by putting fertilisers onto our beds we can actually create lazy plants. This result is a particular shame because plants aren't all that clever; they don't necessarily see the other benefits of having soil life around, such as the organisms' capacity to build healthy soil structure and fight off pests and diseases. Without root exudates, the life in the soil slowly starts to deplete, and over time, the plant will no longer be able to access their services, whether they need it or not.

That's not to say that this process is irreversible, of course. We can effectively 'wean' our plants off fertilisers and build a healthy soil ecosystem. But if you are starting a new garden, you have the opportunity to never get into that cycle in the first place!

There are plenty of other problems with using fertilisers in your garden. (Please note, I do not include pots and containers here as they do not have complete soil ecosystems, so they do need to have nutrients added to them.) The first problem is that, as we already discussed, when nutrients in the soil are water-soluble they can simply be washed

---

Q. I usually use fertiliser on my plants. Should I stop?
A. Yes, you can stop if you are mulching and your plants have access to the soil (meaning, they aren't in containers). Some plants might need to be weaned off fertilisers if you have been using them a lot. All you need to do is mulch your garden and apply fertilisers less and less often over the course of a season, until you stop entirely. If you are just starting out in spring, you can stop completely now. If your plants are in containers, they don't and can't have the soil diversity needed for complete nutrition.

Q. Do chicken pellets and blood, fish and bone meal count as fertilisers?
A. Yes, they do. Anything that adds nutrients directly – that is, in a way that doesn't require anything more than water to make it available to plants – can interrupt the reciprocity cycle between plants and the soil life.

IT ALL STARTS WITH SOIL

away. This means that we can never know how much of what we put in has actually stayed in the soil. When fertilisers are applied on a larger scale, this washing away causes huge problems for our waterways and aquifers, filling them with nitrogen and phosphorous and other things that contaminate and pollute. I'm not suggesting that this would happen in your garden, but at some point you will have a conversation with someone who thinks you're completely bonkers for not using fertilisers. And, if pictures of your gloriously productive and floriferous garden aren't enough, you now have some more argument fodder!

THE RHIZOSPHERE

It's worth taking a breather from nutrients for a moment to discuss the rhizosphere. This is a term you may have come across before, but, if not, here it is. It is the name given to the area immediately surrounding a plant's root, where the majority of exchange between plants and the soil organisms takes place. It's a thriving and busy place. In fact, the rhizosphere, which is usually only a few millimetres wide around each root, can be so densely populated that any unwitting pest or disease-causing microbe will be stopped in its tracks before it makes it to your plant root.

A plant will cultivate a rhizosphere to meet its needs and change it when those needs change by altering the recipe of its root exudates. This adaptability means that plants are always getting the protection and nutrients that they need, just when they need it. It is in the rhizosphere where mycorrhizal fungi are particular showstoppers.

Plants cultivate a rhizosphere to give them the nutrients they need.

MYCORRHIZAL FUNGI

It would be remiss of me not to give fungi their own special shout-out. Because they are absolutely amazing and should be loved and respected for so many reasons. If you are interested in learning about fungi in general, I highly recommend you read Merlin Sheldrake's masterful *Entangled Life*. But for our purposes, we will stick just to what is important for us to know as gardeners.

Mycorrhizal fungi extend in delicate webbed strands beneath the surface of the soil, where they have very close relationships with plants. They connect with a plant's root system, sometimes going as far as actually entering the cells of a plant root so as to form a mutual relationship with the plant. In this relationship the plant gives the fungi carbohydrates, such as sugars. In return, the fungi help the plant to source nutrients and water, extending the reach of the plant's root system by up to 10 times. Fungi can also connect plants to other plants, helping to transfer energy, nutrients and chemical signals.

Top, Sometimes you will see mycorrhizal fungi in compost or pots; it's a wonderful sign that your plants are doing what they're supposed to do. *Bottom,* Sometimes fungi can cause alarm to gardeners, but in almost all cases they are a positive sign of healthy soil.

IT ALL STARTS WITH SOIL

These fungi also create a protective web around the roots of plants: mycorrhiza. The mycorrhizal roots are very delicate but, in layers, provide strong protection from soil-borne bad guys trying to get to your plants' roots. Mycorrhizae are also an important component in creating good soil structure (see the 'Build healthy soil from scratch' section on page 52). In short, mycorrhizal fungi are very important to our plants. This beneficial relationship is why we should not be distressed if we see signs of fungi in our garden and why we should attempt to do what we can to protect these beautiful but incredibly delicate lives.

SOIL pH

Whilst the inorganic components of soil are important and can cause some challenges, most can be overcome by repeatedly adding organic matter and leaving the soil life to do its thing. This simple action will improve soil structure regardless of whether you have fine sandy soils or thick clay soils. However, there is one element of soil that cannot be ignored: pH, or soil acidity. The term pH describes the acidity or alkalinity of your soils, with 0 being the most acidic and 14 being the most alkaline. For your reference, battery acid sits somewhere near pH 0 (most acidic) and drain cleaner sits around pH 14 (most alkaline). Neither of which you want anywhere near your kids! Right in the middle, at around pH 6.5 to 7, is neutral. Pure water is neutral. Most plants are happy in neutral soils, with the exception of certain plants that thrive in slightly acidic soils.

These plants include blueberries, azaleas, rhododendrons and acers. Whilst they will grow in alkaline soils, they won't perform to their best. Blueberries in alkaline soils, or even neutral, will struggle to produce fruits, for example.

The first signs of high or low pH will become apparent when looking at plants that grow well in the local area. A vibrant cornflower-blue hydrangea is a sure sign of slightly acidic soils (or a very attentive gardener), as are blueberry bushes groaning under the weight of fruit. Alkaline soils are a little more challenging to spot, but usually these occur in chalky areas. Most of the signs are subtle, and the changes in soil pH will also be pretty subtle. But chances are you can grow what you like in either soil without noticing much difference.

The problems come when it gets a bit more extreme (which is uncommon). Soil with a pH of 7.8 or higher will be lacking in available copper, zinc, manganese, phosphorous and iron. This deficiency occurs because at this pH these nutrients are in solid form and are, therefore, more difficult for plants to access. Certain plants can cope without these nutrients well and are adapted to it, but others can develop a condition called *chlorosis*, which presents as pale leaves with dark veins. The paleness occurs because the nutrients listed above are needed to create chlorophyll (the green colour) in leaves, which powers photosynthesis. Many people see this condition in their plants and assume it is because of a nutrient deficiency – which it is to some extent. But adding nutrients to the soil will not help, because the added nutrients will change to their solid form due to the pH. Whereas fixing the pH will make the existing nutrients become available. However, it should be noted that life in the soil will help unlock these nutrients regardless of the pH. So, if you have soils rich in organic matter and full of life, your plants are not likely to suffer.

What is far more likely to be a problem is when we add anything to our soil without adequately checking them first. Adding an amendment such as lime or sulphur is needed only in extreme cases and should be done with great caution. Otherwise, you could create pH problems in your soil where there weren't any before. Please be aware that it takes months to adjust the pH of your soil, so, when adding any amendment, make a modest application, wait a few months and retest before applying any more.

If your garden is suffering from high pH due to salt or overuse of fertilisers, you're better off trying to improve the drainage than trying to fix the pH. Why? Because regular rainfall gradually leaches away soluble alkalinity. This process is called *acidification* (making something more acidic), and it takes place naturally in warm, moist environments.

Where I live in Devon, the soil tends to be slightly acidic simply because the climate is very wet, which means that over time the ground acidifies. Where the soil is sandy and quick draining, it is much more acidic than where it is clay soil and the drainage isn't so great. There are some additives you can use if your soil pH is way off, so take a look at the 'Soil amendments' section on page 57.

IT ALL STARTS WITH SOIL

Left, A soil pH meter is an easy, reusable way to test your soil. *Right*, A soil testing kit is usually more accurate, but the results can be difficult to read.

To test your soil pH

There are several different ways to test soil pH, depending on the accuracy you need and how much you want to spend.

Soil pH meter. The simplest test is using a soil pH meter, which is a little display on the top of a long metal spike. They're relatively inexpensive and straightforward to use. Helpfully, most of them also include a moisture meter, which is also very useful. The spike is inserted into the soil and the display shows you clearly what the pH is. Use this several times, changing the location and depth slightly each time in between to get a more accurate reading.

Soil testing kit. These kits come in several shapes and sizes, but the simplest and most inexpensive is a small set of materials with a test tube or two inside. Each tube contains a sprinkle of something roughly resembling sand. You add a little bit of soil and some filtered or boiled water and shake. Once the sediment has settled,

GROW A NEW GARDEN

A DIY soil test is easy and can show some interesting results.

you will be left with coloured water. Check the colour against the chart to check your pH. You may want to take small samples from different depths and locations for a more accurate overview of the pH of the garden.

DIY soil test. If you haven't got a soil testing kit to hand, take two small bowls and put in each 1 spoonful of soil taken from about 7 to 10cm (3–4in) below the surface. If your soil is dry add a little bit of filtered or boiled water to the soil to make a paste. Then add half a cup of vinegar to one of the bowls and mix gently. If it fizzes, you have slightly alkaline soils. In the other bowl, add half a cup of bicarbonate of soda and mix. If the mixture fizzes, you have acidic soil.

The neighbourhood snoop. The cheapest option: check out what is growing in your neighbourhood. If your neighbours have a great crop of blueberries, you might be dealing with acidic soil. If you have wonderful calcareous grasslands, full of wildflowers and

These gorgeous hydrangeas definitely tell me the soil here is acidic, but variations can happen even within a single garden.

orchids, its more likely to be alkaline. If the hydrangeas are a vibrant cornflower blue, the soil is acidic; if they're coming out pink, it's alkaline; and if they're purple, it's pretty neutral. Whilst this method isn't completely fail-safe, it's a great place to start.

ASSESS YOUR SOIL

Conventional soil assessments will tell you to figure out if your soil is clayey or sandy. This is pretty basic, and most of us probably know the answer quite instinctively – particularly if you have extremely sandy or clayey soil. If it's light and grainy, it's sandy. If it's hard when dry and slippery and smooth when wet, it's clay. If it's somewhere in between and looks like 'proper soil', then you likely have some sort of loam. Loam is the so-called 'ideal' soil, but it is also quite rare. It's a spectrum, too, from sandy loam to clayey loam.

If you still aren't sure what kind of soil you have, you can perform this simple test. Take a handful of your soil and squeeze it. If it sticks

Left, If your soil can be rolled into a ball, it has a high clay content. My soil includes Devon red clay, so the colour is very dramatic! *Centre,* Squeezing the ball of soil can tell you about the organic matter content. *Right,* If the soil ball holds its shape well, there is poor aggregation and not enough organic matter.

IT ALL STARTS WITH SOIL

together, it is either loamy or clay; if it doesn't, it is sandy soil. If it stuck together, take the soil ball and squeeze it gently between your thumb and forefinger. If it falls apart, you have loam (lucky you!), but if it stays together and simply moulds to your fingers, you have clay. If your clay or sandy soil contains plenty of organic matter, it may crumble apart easily (in the case of clay) or stick together to some extent (in the case of sand). Soils with plenty of organic matter will generally appear a rich, dark brown colour.

Personally, I don't find it particularly useful to dwell too much on the type of soil you have when it comes to soil care. Particularly as we can't change the soil we are working with. It is useful to know for growing certain plants, though. For example, don't try growing plants that love free-draining, slightly acidic soil if you have heavy clay soil – this is what pots are for!

When I first moved to Devon, I was desperate for some space to grow in. I had a small back garden that I turned into a vegetable patch, but I had been used to a large back garden and a polytunnel. So I asked around, and soon my cousin approached me, asking if I would like to share my aunt's polytunnel. It was a few miles from where I lived, but absolutely perfect. My late uncle had procured it for their smallholding around 15 years before, and it was huge – around 27m (90ft) long. There was plenty of space for my auntie and me to share and far too much for her on her own. It was, after all, my uncle who had been the gardening enthusiast. The soil there was so sandy that you could scrape off the top layer of soil and reveal bright yellow sand underneath. There were no earthworms to be seen. When I watered, it would run off and pool at the lowest point, barely soaking into the ground at all. But plants grew in it just fine – with the help of some fertiliser. Back home in my garden, the soil was heavy clay. It was difficult to get a spade into, and in winter it was wet and slimy. Whilst these two soils could not have been more different from one another, I treated them in almost exactly the same way. By adding organic matter.

Now, I don't want to admit to being lazy, but I find it difficult to contemplate how tiring it would've been to dig over the soil in my garden. Moreover, my landlord might have had something to say about it. On the opposite end of the scale, if I had dug the soil at my auntie's polytunnel and there happened to be a stiff breeze, the soil would have simply flown away and decorated the nearby hedgerow.

IT ALL STARTS WITH SOIL

That being said, my auntie and, before her, my uncle had been digging on this site for many years with great success. But as we've already seen, the soil was quite severely depleted, and the crops needed a strict fertiliser regime to keep them cropping well. I only used this polytunnel for two years, and all I did was take some compost and lay it on the surface of the soil once a year. I used as much as I could; I would estimate 3 to 5cm (1–2 in). Ideally, I'd have used more at first. My auntie used around the same amount but dug it into the soil. This practice is designed to put the nutrients within the compost directly where the plant roots are. However, by applying the compost to the surface of the soil, we can employ the life in the soil to drag the compost down to where the plant roots are and bring the nutrients as they go. Additionally, we end up with a nice surface that helps the soil absorb water. As soon as I applied a layer of compost to the top of the soil, water no longer ran off the surface and pooled at the lowest point. It was simply absorbed into the soil.

When I put my finger just below the surface of the compost, there was a layer of moisture trapped underneath, even though I was growing in depleted soils that apparently did not have any earthworms in them. In total contrast, the beds that belong to my auntie were still dry on top and needed a lot more watering. I grew a very successful crop of tomatoes, chilies, aubergine, flowers, cucumbers, courgettes and pumpkins. And the best part was, I didn't fertilise once. Straight away the soil life kicked into gear, with the compost acting as an inoculant, adding life to the soil. After two years, the difference in soil on the two different beds was incredible. And worms even returned to my side.

My auntie also lent me her vegetable patch. Did I mention my family is amazing? We divided it down the middle. Her side was rotovated, and mine received a tractor full of manure from a local farmer. The manure certainly had its challenges. It was clumpy in places and contained a lot of turf netting that ended up tying large chunks of manure to other large chunks of manure. It also contained impressive stones, the likes of which I could never have transported in the back of my car but evidently the tractor had no trouble with. However, it did contain a large number of earthworms. With plenty to

Adding organic matter is key to improving most soils.

GROW A NEW GARDEN

These two samples were taken a few inches apart. The soil on the left hadn't been dug for two years and had compost layered on top; the soil on the right had been regularly dug to incorporate the compost. Both had the same amount of organic matter added. The difference in organic matter content even a few centimetres down is incredible.

eat, the earthworms set about incorporating the manure into the soil. And once again my crops thrived with very little work for me. In fact, I even managed to grow two giant pumpkins with no fertiliser at all, just a large helping of manure.

I'm a scientist, and this case study certainly would not stand up to peer reviews, but the results do speak for themselves, and it's important to remember that my method took a lot less work. My auntie even had to enlist the help of my strapping, young cousin to do the rotovating because, even using a purpose-built machine, it is really hard work.

IT ALL STARTS WITH SOIL

What most soils are lacking is organic matter, so here we will do a basic assessment of our soil to make sure that we are not working with any extremes, to find out what we can expect to plant.

Basic soil structure assessment

In carrying out this assessment, we are looking for signs of aggregation. Aggregation is soil particles being stuck together by organic glues. Soils that are aggregated will absorb and hold on to water more easily, have better drainage and generally be in better health. The more aggregation you have, usually the more organic matter the soil contains. For more information on aggregation, check out the 'Build healthy soil from scratch' section on page 52.

To check the aggregation of your soil you will need to perform the soil test described on page 44. Soils that present as very clayey or very sandy during this test will be lacking in organic matter and therefore aggregation. When there is a good amount of organic matter, clay soils will fall apart easier and sandy soils will stick together easier.

You can also perform an infiltration test on your soil. This is easier than it sounds and only involves a bottle of water. Fill the bottle to the top and quickly turn it upside down on top of your soil, pressing it to the surface as much as possible to ensure water doesn't leak out of the sides. If the water stays in the bottle, you likely have clay soil or compacted soil without much aggregation. If the water drains out of the bottle quickly, this indicates you have good aggregation or particularly sandy soil (you should know which from your earlier soil test). This simple test will give you an idea of how much of the water landing on your garden is actually going into your soil and reaching your plants' roots.

Visual soil assessment

It worth putting your hands into the soil and getting a feel for it, but there is a lot that can be told about your soil just from looking at it. Whilst soils are different colours all over the world, there are a few rules that will apply universally. Dark-coloured soil is generally higher in organic matter. Peat, which is almost completely black, is almost pure carbon (organic matter) created from decomposing peat moss. If the soil in your garden looks pale compared with other soils in the area where you live, it's probably in need of some serious TLC.

GROW A NEW GARDEN

Left, Peat is almost pure carbon and has a very dark colour. *Right,* My soil is poorly aggregated clay, which is obvious in winter when it is wet and sticky and in summer when it dries and cracks.

You may be able to get a feel for the depth of your topsoil just by looking at it. If you see rocks protruding through the surface, the topsoil is likely quite thin. This often happens in highland and coastal areas. There are several ways to deal with this kind of environment that we will look at in the section 'Your dream garden' on page 73. It may be obvious to you what kind of soil you have if you live on the edge of a desert, near a cliff, by some sand dunes or on the edge of a wetland. But you would be surprised how many people don't take this into account when planning their garden.

BUILD HEALTHY SOIL FROM SCRATCH

Let's make one thing perfectly clear. When it comes to your soil, you can improve what you have, but you cannot fundamentally change it. Adding different constituents to your soil will not turn it into a different type of soil. In order to use what you have, you must understand what you are working with and plant appropriately. This means using plants that are suited for your environment. You can also use containers and raised beds to create areas where the soil is completely different. This helps to raise the profile of the soil and gives you the ability to control the soil environment much more accurately.

IT ALL STARTS WITH SOIL

If you had really thick clay in your last garden and have really thin sandy or silty soil in your new garden, it would be tempting to think that you needed to treat these soils differently. Or that these two soils need different things to correct them. However, these two types of soil have one very important similarity. They are both lacking aggregation.

It makes sense that you might want soil particles sticking together on sandy soil. If you try to dig it on a particularly windy day, you risk losing half of it to the nearest hedgerow. So, having it stuck together a little sounds like a good plan. In clay soil, where the particles are firmly pressed together, desiring aggregation might not make sense. Clay particles are like tiny little plates, and they sit on top of one another, nicely tessellating. This doesn't allow for much water to flow down through them, so water gets trapped in the top layers. That is why, when we step on wet clay, it becomes slippery. The plates simply slide over one another, and the net result is we end up with our bottoms in the mud.

So why on Earth would we want these particles to be more stuck together? Let me explain. Clay expands when it gets wet (making that sticky gloop) and then contracts when it dries, causing cracking. This happens when all the particles in the soil are equally stuck to one another. If we have some particles that are more stuck to the ones around them than others, the effects of expanding (becoming gloopy) and contracting (cracking) are lessened. When the particles in the soil are stuck to one another, they move away from other particles, creating spaces called *pores*. These pores allow water to travel down through the soil rather than gathering at the surface. This drainage helps the soil to not become saturated, and helps it not dry out as quickly because water that isn't so close to the surface isn't so easily evaporated.

In addition to creating drainage, aggregated soil also holds water well. Water clings to the surface of each aggregate, coating it. The slight chemical charge on an aggregate helps the water to stay put and therefore aggregated soils are less prone to drying out, and less prone to flooding due to complete saturation.

The clever thing is that most soils become aggregated all by themselves. So why do we end up with thick, gloopy clay soil that makes us slip and fall? Or soil so sandy that we have to water it twice a day? These things occur mostly in disturbed or depleted soils. Think of a muddy building site or a well-walked dirt path. Sometimes our

GROW A NEW GARDEN

Garden soils can be amongst the most heavily disturbed of soils, creating problems for plants.

gardens and allotments can be at the top of the list of disturbed and depleted soils.

To understand why disturbing soil can lead to depletion and poor soil structure, we must first understand how aggregates and good soil structure are formed. The soil is a teeming ecosystem full of tiny organisms interacting with one another and moving about. Bacteria are amongst the smallest and most abundant of these organisms (you could fit thousands of them just on the full stop at the end of this sentence). As bacteria move through the soil, they secrete tiny amounts of sticky sugars, called *polysaccharides*. These sticky sugars bind the tiniest particles in the soil into slightly less tiny particles. Binding at this scale is called *micro-aggregation*. Fungi in the soil also produce a sticky substance, called *glomalin*. This substance binds those slightly less-tiny particles together further. Fungal hyphae – the tiny, delicate, hair-like filaments of fungi that make a web beneath the soil – also entangle these particles, holding them

IT ALL STARTS WITH SOIL

The crumbly texture of soil is built with organic glues that cause aggregation.

together even more and forming full-fledged aggregates. We gardeners refer to these as *peds*. It's what gives us that beautiful, crumbly texture that we so desire.

When we dig our soils, we break the fungal hyphae and cause the breakdown of the bacterial glues. This means that we lose the aggregation, and with it we lose the pores in the soil. We also lose some of the

Q. My soil has been regularly dug for many years. Will it be OK?
A. Yes, your soil will be fine. Regular digging isn't good news, but soil can regenerate really quickly when it is allowed to. Depending on the size of the area you are working with, your soil should recover in one to two years with the application of organic matter, and continue to improve from there.

GROW A NEW GARDEN

life in the soil, which had been helping to feed nutrients to our plants and keep them pest- and disease-free.

We humans are natural problem-solvers. Typically, when it comes to our gardens and allotments, we see heavy clay soil as a problem that needs solving. Often, our solution is hours of back-breaking work digging it, turning it over and adding organic matter. The end result is usually that nice crumbly texture. But, by the same time next year, we are back to where we started. It becomes a problem again, and we fix it the same way. Similarly, we see sandy soil as a problem and may try to dig in more organic matter to fix it, which can actually exacerbate the problem.

If we let soil build itself, the result is a structure that improves year after year. In fact, naturally built soil structure is so stable that it can

Left, Organic matter was the first thing I added to my new garden, and on the lawn I over-seeded, too. *Right*, On my beds, I added a deep layer of organic matter. This meant I could plant straight into the beds. After just one year, the soil was much improved.

IT ALL STARTS WITH SOIL

> **Q. How do I plant something without disturbing the soil?**
> A. Don't worry about digging holes. There is no way that you can plant most things without first digging a hole. Soil problems start to develop when widespread, repeated and prolonged disturbance happens, such as digging an entire bed over annually.
>
> **Q. Does no-dig gardening apply to growing flowers, too?**
> A. Yes! Soils can be treated in more or less the same for a wide number of plants. Some plants will 'mulch themselves' by dropping leaves, and you needn't do anything to the soil beneath these. We use no-dig gardening primarily when growing smaller and more fragile plants. Vegetables and flowers both benefit enormously from healthy soil.

withstand heavy machinery, flooding, drying out and strong winds, which is a lot more than can be said for most garden soils, especially in newly built houses.

So, whether your allotment is heavy clay or light, dusty sand, nature has the answer. All we need to do is let it get on and do its thing. In nature, dead organic matter, like leaves, falls to the surface of the soil. This is where the organisms in the soil will go to feed. So, instead of digging in your organic matter, lay it on top like nature intended and watch your soil become healthier and more resilient year after year!

SOIL AMENDMENTS

There are several materials that we can add to our soil that will help to improve it. Some of these materials are better than others. Here, we will take a look at some of these products, considering which ones are best and why.

Topsoil

Topsoil is the name given to the top few centimetres (or about 1 inch) of soil, the part where most of the biological activity takes place. Earlier in this chapter, we discussed how soil structure is created. If you live in an area that doesn't have much topsoil, either because of the area's geography or because of how your garden or allotment was created, it can be pretty tempting to buy in this topsoil. Additionally, if you live somewhere where the soil conditions are not favourable, you can buy in topsoil and essentially create a blank canvas on which to work. But the problem is that, even though the inorganic components are the same as the topsoil we want, the nature of 'topsoil product' is entirely different.

Sometimes topsoil is dug out from somewhere and bagged up to sell. This often happens where it is removed to aid building work. During the process, any organic glues that were present will have broken down. As soon as the organic glues encounter air and UV light, they will oxidise, turn from carbon into carbon dioxide and return to the atmosphere. The tiny lives within soil don't take kindly to their homes being extracted, screened or bagged, so commercial topsoil tends to be devoid of life, except weed seeds that are almost impossible to remove from it. Some topsoil contains perennial weeds that grow from fragments of plant matter that wasn't removed. These types of weeds, such as Japanese knotweed, can be extremely difficult to remove from your garden and can even devalue your house. So, it's always best to avoid introducing them if possible.

Many topsoil products are made simply by blending ingredients. The ingredients are sourced from various locations, including building sites and quarries, and are then screened. This essentially divides it according to particle size – we might more readily call it sieving. The particles are blended in a way that roughly resembles loam – 20 per cent clay, 40 per cent silt and 40 per cent sand. Then a nice big helping of organic matter is added to ensure that there are some nutrients for our plants. And the whole thing is bagged up.

This manufactured topsoil does not do a great job of adding life to the soil and adds hardly any organic matter. If you are lacking topsoil, what you are most in need of to build it back up is life. If you don't have enough organic matter – which many topsoil products do not – there will not be food for the life in the soil, and any soil organisms simply won't be able to thrive. Moreover, the organic matter needs to be on the surface for it to be available to our friendly earthworms, who, whilst great at digging, are not clever. They go up and down and very little else. So, they need organic matter at the surface to help

IT ALL STARTS WITH SOIL

I used topsoil to level part of my garden that gathers water.

them find it. However, if you need to fill a hole or level an area, topsoil will do the job much better than straight organic matter because it will not be eaten and taken into the soil.

Sand

Sand is often recommended as a way to help improve drainage in heavy clay soils. The idea being that, if a soil is too clayey, adding sand will help to balance the mixture of ingredients and turn it into something more closely resembling loam. The problem is, if you have a soil that is 60 per cent clay, you will have to add an entire beach to it to be able to tip the balance. Moreover, without aggregation of the particles, the clay will just filter into the spaces between the particles of sand.

Think of it this way: When you're making cement, you add a lot of clay. Unless you use a lot more sand than you should, it doesn't make the cement any softer. The way I think of this is like a crumble topping. If you try mixing sugar and flour (sand and clay), you will never be able to get that nice crumbly texture that is light, fluffy and full of air. As soon as you add butter (organic matter) and start to mix it together, the particles will start to stick together (to aggregate), and the entire mixture will grow as more air is incorporated. This has quite a similar structure and feel to well-aerated soil, and neither could have this structure without particles being stuck together, that is, well-aggregated.

GROW A NEW GARDEN

It is for this reason that I don't recommend using sand in your garden: it doesn't create better aggregation. In my professional opinion, sand is not worth your time or money.

Lime

Many people will recommend using lime, especially where the soil is clay and not well aggregated. It may be necessary to add lime to your garden, but never add it without properly testing your soil's pH first. If your soil is particularly acidic, adding some lime can help to balance it, neutralise it, and create a healthier soil environment. However, lime should be added gradually. Retest every 6 to 12 months and add a small amount more if needed; roughly half a cup per square metre, sprinkled lightly over the surface. Ordinarily, soils become acidic when they are being leached. This occurs when there is too much water passing through them, and the nutrients are being washed away. If this is the case, you should try to address the underlying problem as well as fixing the pH.

Only use lime sparingly, and only when you have confirmed it is needed.

IT ALL STARTS WITH SOIL

Gypsum

Gypsum helps clay particles bind together, but only if the soils were slightly salty (sodic) to begin with. If you live in a coastal area and your garden experiences spray from the ocean or regular sea breezes, your soil is likely to be a little salty, and gypsum may help. Salt will cause the clay particles to become loose, and so they will fill in all the tiny spaces in the soil. Adding gypsum will help to dislodge the salt from the clay particles and can help aggregation. It's not a good fix if you don't have potentially sodic soils, though.

Compost

This is by far the best thing you can add to your soil. Compost is decomposed organic matter, either plant waste or manure, that has had enough time to break down and rot so that it resembles soil. The great thing about adding compost to your garden is that it feeds the life in the soil, which then, in turn, improves the soil structure, releasing nutrients to our plants and creating a healthier soil environment. You can also use compost in containers. If it is a good-quality compost, it will support your plants' needs throughout the growing seasons,

Top, Gypsum can help if your clay soil is slightly salty but is otherwise not needed. *Bottom,* Most soils are lacking in organic matter, so adding compost will help to improve this, having a knock-on effect on drainage, absorbency and fertility.

GROW A NEW GARDEN

> **Q. I have been adding organic matter to my garden but the drainage hasn't improved. What's happening?**
> A. You may need to look for the source of your drainage problems and consider adding some additional drainage. For example, if you have a drainpipe emptying into your garden, you can act to divert this, either to a water butt or a soakaway. If you have a large area of patio that gently slopes towards your lawn, it might be catching water and tipping it towards your garden. You can add some drainage between the patio and the lawn or divert the water flow somewhere else. You may want to consider looking at sustainable urban drainage systems. These are neat and eco-friendly solutions to addressing excess water.

though you will need to top it up or change it completely at the end of the year. I recommend adding it back onto your compost heap at that point or using it as a mulch.

Obtaining compost can be a bit of a minefield. There are so many different types, it can be difficult to know which one will be right for your garden, or your particular set of circumstances.

SOIL IMPROVERS

These are rich composts packed with nutrients. Don't use this for young plants or seeds because the high nutrient value will be too much for them and can produce abnormal and unhealthy growth.

Farmyard manure

This source is great and can also be sourced for free relatively easily, but it can come with a host of contaminants. If the farmer is conscientious about the use of wormers in their farm animals, the manure will probably be OK because little to none will pass through the animals. However, if they use a lot of wormer – like they do in particularly intensive farms – this is probably not good news for your soil. A good test is to see whether the manure itself has worms in it. If it's healthy, it will probably have a lot!

IT ALL STARTS WITH SOIL

Testing for weed killers

If you're ever unsure about a manure source, a good test for weed killers is to grow some broad beans. Before spreading it on your garden, make up a pot of the manure you intend to grow in. Plant a small broad bean plant into this manure and then wait a few weeks. If you have a beautiful, healthy-looking plant, the manure is probably OK. If the top leaves start to twist, bend or look stunted, this change indicates herbicide poisoning. You should not use the manure under any circumstances because it will affect your plants.

The twisted leaves of a tomato plant being poisoned from weed killer in the soil.

IT ALL STARTS WITH SOIL

Good compost is dark in colour and soft in texture.

Also watch out for farmyard manure that could contain broadleaf weed killers, such as aminopyralid. These kill the plants but don't harm the animal; instead, they pass straight through, into the manure. These types of weed killers can contaminate your soils for years. So, it's a good idea to talk with the farmer, or stick with a manure source that is already used and loved by other gardeners. This will require you to ask around in your local area. Your friendly neighbourhood gardening club will certainly be able to point you in the right direction.

Green waste compost

Green waste is unwanted organic matter and is usually obtained from your local waste disposal centre. This unwanted matter is usually from people's gardens. Whilst it can be a good resource, the quality variance is enormous. Where I live, the green waste is of a really high

You may see fungal mycelium in compost, sometimes, which is a good sign that the compost is healthy.

Table 2.1. Good versus bad features of bought-in composts

ATTRIBUTE	GOOD	BAD
Colour	Very dark in colour – a dark brown, almost black. There might be flecks of white in the compost and this is not something you should worry about. It is often a sign of healthy compost and the presence of fungi.	Pale or tan-coloured. This indicates a less-fertile compost, but it may still be used as a mulch to add structure to the soil.
Consistency	Relatively even.	Large lumps. Whilst lumps can be broken down manually by hand, their presence in the first place indicates that the compost may not have been stored correctly or is just not of a high quality.
Smell	Little to no scent, beyond 'dirt'.	If you open a bag of compost and it smells really bad, this is usually a sign of anaerobic bacteria. These bacteria exist where there isn't much, or any, air, so the compost could be too wet. You can try to dry it out or add it to your compost pile. Otherwise, the compost could still be in the early stages of decomposition. This kind of compost works as a mulch or soil improver but should be kept away from young plants and seedlings.
Presence of little pellets	No little pellets means no synthetic fertilisers, which is best for fostering soil health.	It is common to find small pellets in your compost. They are usually pale yellow in colour and pop if you squeeze them. These are often mistaken for snail or slug eggs, but they are, in fact, fertiliser pellets. They are added to many composts as a slow-release fertiliser with a coating that breaks down gradually inside the soil. Some of these pellets are bright blue. They are nothing to worry about, but if you are interested in fostering healthy soil you should avoid adding these kinds of compost to your beds, as they will interrupt the nutrients and carbon cycling between the soil and your plants. If you find the pellets, use the compost in pots or containers instead.

IT ALL STARTS WITH SOIL

quality, but in some areas it's terrible. I recommend asking around to see if anybody else has used it before committing to using it yourself. If you are an experienced gardener, you might be able to tell the difference already between good and bad compost, so you could pick up some and try it. If you are not an experienced gardener, please understand that some composts are better than others, and if your seeds don't germinate or your plants don't thrive, it is sometimes the fault of poor compost.

These are snail eggs; slug eggs look similar, but they do not cluster like this.

Homemade compost

Homemade compost is by far the best resource that you could possibly have in your garden. It's teeming with life. If your garden is organic, it will be organic, too. And the nutrients in the compost will be exactly the nutrients that your plants need during their development in your garden. In other words, it's a tailor-made product that perfectly caters to the unique needs of the plants in your garden. If you're starting a new garden, chances are you don't have any homemade compost just yet, but you should set up a composting area as soon as possible to start making use of this wonderful resource. For a head start, go to 'Making your own compost' in Chapter 7, page 165.

CHAPTER 3

Plan and design your garden

Now that you're a total expert on getting your soil, you should have no trouble creating a healthy garden that thrives year on year. So, let's move on to creating your garden. Grab a pencil and a sheet of paper and let's go. Planning your garden is a fun and creative task that allows your imagination to run riot.

BASICS OF GARDEN DESIGN

Every time I take on a new garden, I pile a lot of expectations on myself. I want to achieve something that is beautiful, functional and 'technically correct'. The only trouble is that I'm not a garden designer. So apart from a sound knowledge of plants and their individual needs, I'm not great at this. I will admit that when I moved into my new house I considered taking a garden design course. This consideration was short-lived as, upon doing some research, I realised how much it would cost me in both money and time. So, instead, I did what any other self-respecting gardener would do in my position and picked up some books and attempted to teach myself garden design. How hard could it be? It turns out that, as with most subjects, the more you know, the more you know you don't know. As the Alexander Pope line inscribed on the fountain in one of my favourite gardens, in West

West Green House with the fountain in the background: garden design like this is an art.

GROW A NEW GARDEN

Green House, Hampshire, goes: 'A little learning is a dang'rous thing; drink deep, or taste not the Pierian spring.'

Garden design, of course, is nothing like interior design. One cannot simply roll out a design just like you saw on your favourite TV programme and expect it to work like you can with your living room. Garden design must take into account the specific conditions in the specific garden. This includes but is not limited to:

- the soil
- the prevailing wind direction
- the direction of the sun
- pre-existing features
- whether your neighbours intend to prune their hedge regularly
- where the dog usually 'goes pee pee'
- whether or not plants like heuchera 'like the look of the place'

It has to take into account the requirements of the plants being used, and it has to adapt through the seasons. Garden design is incredibly nuanced and requires a lot of knowledge about plants.

It's worth noting at this point, before we run screaming from our gardens, that it doesn't really matter that much if you get it wrong. It's more cost-effective to get it right but, even with the best of intentions and knowledge, sometimes things just don't work out in gardens. Happily, most of us will never have a team of judges coming to inspect the exact positioning of our azaleas or critique our use of colour. The pressure comes only from within. So, if you're worried about getting started, like I was, just accept that this will be a journey. It will be a process of learning about gardening and about your garden, and things will go wrong along the way.

Never compare your garden to a show garden, ever. It's tempting; you see something you love, and you want to recreate that in your space. But behind every show garden is a team of expert horticulturalists growing and selecting the perfect plants for the display. Flowers are tied up, so they will be in bloom at the exact right moment. Before the crowds appear, plants that are looking a little lifeless or have been chewed on in the night are replaced by backup plants and anything even slightly unsightly is removed. And, at the end of the show, the whole garden is cleared away. It lasts two weeks, maximum.

PLAN AND DESIGN YOUR GARDEN

Show gardens are just that, for show. You can't compare your garden to them, as your garden is permanent. (Addleshaw Goddard Junglette Garden, Chelsea Flower Show 2024, balcony gardens gold-medal winner).

Your garden, however, is a living space, beautiful in its dynamics and actuality. It might look fabulous a week before your garden party and then be bulldozed by high winds the night before. It will grow and change during its life. Plants will grow, trees will cast shade, some plants will die, pests might arrive en masse, and curious garden ornaments will be acquired (if that's your thing). Your own tastes might change, or your needs might change. You may acquire a lovable dog with a penchant for digging holes. Your elderly mother-in-law may come and stay with you and demand that you make your garden fully accessible. Like me, you might have children, which means you absolutely must have a lawn so that they can run around and do cartwheels. Or, as we saw in 2020, our priorities could change overnight, meaning that the most important thing in our world is now hoarding toilet rolls and growing tomatoes.

The point is that a garden needs to be dynamic, and achieving perfection alongside all the challenges and parameters is simply not possible. So, I invite you to leave your perfectionism inside and step out into the great unknown. Part of the magic of a garden is never quite knowing

GROW A NEW GARDEN

Top, My current garden when we first moved in. *Bottom*, Kenny is one of my favourite people and a font of gardening knowledge.

what you will find. One day, it might be an unexpected flower, and the next, a seemingly biblical plague of caterpillars eating everything in sight.

The only trouble is that not only am I not a garden designer, I don't aspire to be. I don't have 'the eye' for great plant pairings, but I can tell when something looks nice or if it looks awful. Given that my garden is under constant scrutiny from the online community and my various horticultural friends, the pressure on me feels like it is outwards, as well as inwards. But in real life, unless you have a dedicated and highly skilled team, no garden is perfect.

When I moved into my current home, I really wanted to get it right. The hard landscaping was done by the previous owners, but, except for three large birch trees and some box plants, the garden was devoid of plant life. Finally, I had a small budget to play with, instead of having to scrounge plants from anywhere I could.

Fortunately for me, I have lots of wonderful connections from my years pretending to fit in with the horticultural industry. One of my favourites of these connections is my now-good friend Kenny Raybould. Aside from

PLAN AND DESIGN YOUR GARDEN

being one of the nicest people I know and a fellow lover of gin, Kenny is a fabulously knowledgeable plantsman, garden designer and florist. What he doesn't know about garden design simply isn't worth knowing. So naturally, I asked him for some advice. Luckily, Kenny was happy to distil his years of experience into a few easy-to-follow pieces of advice, which I am delighted to be able to pass on to you in this chapter.

YOUR DREAM GARDEN

OK, it's time to grab yourself a nice cup of tea, and let your imagination go wild. Picture yourself in your future garden – not the one you will have after winning the lottery, the one can you have now, but when it's finished. We need to start with how you will use your garden. This is closely tied to the reason why you want to create a garden in the first place. If you want, you can write down both. Don't hold back; you can always scrub out unrealistic things later.

Your wants

Start by making a list of all of the things that you want to do in your garden. This can include but is certainly not limited to:

- sunbathing
- dining al fresco
- drinking coffee in a warm greenhouse
- playing football with your kids
- showing off your new pizza oven
- hosting barbecues

I love warm summer evenings in the garden and growing my own food.

GROW A NEW GARDEN

- setting up a pond-plant nursery
- having quiet time with a book
- swimming
- hosting open-air cinema nights
- just messing around with plants

You might be particularly adventurous and fancy trying all of the above, and that is absolutely commendable. But more than likely, there would just be a few things that you can really picture using your garden for. At this point you should probably consult with anybody else who will be likely to use the garden. Before designing our garden, I spoke with my partner, and he dreamed of evenings outside by a warm fire with a glass of beer. The kids wanted to do cartwheels and make daisy chains (no prizes for guessing that we have girls). And we didn't ask but were fairly confident that the dog just wanted somewhere to bark at all of the neighbourhood cats. Whilst it might not be possible to cater to all of your most fantastical whims, creating a list like this will give you a good idea of what you want to work towards. And once you have finished with the garden, you'll look back on this list and smile, hopefully whilst doing one of the things that at one point you could only dream of.

The theme

The next thing that we're going to do is think about theme or vibe in your garden. This is where we are going to use our imaginations, Pinterest boards and Instagram accounts. It will pay dividends to take some time to visit local gardens, too. This will give you an idea of the kinds of plants that grow well in your area and will give you some inspiration. Spending some time in different gardens can help you to learn what you do and don't like. When deciding on a theme or a vibe for your garden, you can base this around either a particular style of garden or some of your favourite plants.

If you're already a gardener, it's likely that you have a few favourite plants. I am absolutely in love with *Rudbeckia* 'Goldsturm' and *Salvia* 'Amistad'. The first is a yellow daisy-like flower with a black centre, and the second is a tall purple-flowering plant with long spikes of trumpet-like flowers. They're like old friends to me. I see them and they make me happy. Yellow and purple are quite a bold combination,

Left, A formal garden centring around water. *Right,* A tropical garden full of huge foliage.

and they are not ones that most garden designers would advocate. But this is my garden, and I am making it for me (and my family), so I choose to welcome both of these beautiful friends simply because I can. You can do the same in your garden. If there are some plants that you really want to grow, you can base the rest of your garden design around those. If you're not sure what plants you like, you might prefer to stick to a theme or style of garden that you particularly like. I recommend choosing something that ties in with the local environment. For example, if you live near the coast consider creating a coastal garden, or at least using some plants that grow locally to make your garden seem more natural.

Here are some examples of garden themes/styles:

Mediterranean. Incorporate soft hues, gentle grasses, olive or cypress trees, and terracotta or stone colours.
Tropical. To create a jungle feel, use large foliage and bold colours, focusing on height and depth.

GROW A NEW GARDEN

Cottage garden full of wildflowers and colour.

Coastal. Use gravel, sand, shells and other hard landscaping along with some bright flowering plants and lots of texture. You can also use things like driftwood, rope and other nautical trimmings.

Cottage. Focus on traditional planting with lots of colour and deep borders. Add in some vintage furniture.

Contemporary. Use crisp lines, minimalism and modern furnishings to create a fuss-free and smart-looking space. The emphasis is on foliage when it comes to planting, and not too many flowers.

Formal. Use straight lines, repeating patterns and, if possible, incorporate a centrepiece, such as a fountain or seating area. Planting is usually light colours, with an emphasis on white.

Wildlife. Focus on encouraging biodiversity. These gardens often play home to several ponds, a log pile, long grass, wildflowers and a bug hotel.

Vegetable. Focus on growing vegetables and flowers: beds overflowing with produce, a greenhouse to raise seedlings and, more often than not, some fruit trees.

If you're anything like me, you don't necessarily want to go with just one of these themes. My own personal style is lots of colour, a bit of chaos, overflowing beds and lots of fun, with some nice wildlife-friendly areas and somewhere for me to lie down at the end of a long

My garden in its first summer.

GROW A NEW GARDEN

day. I like to grow some vegetables in my garden – salads in particular – and if a tree wants to give me some fruit as well as some shade, I'm definitely here for it!

It is entirely possible to take elements from each of the previous themes and create your own homespun theme that suits you perfectly. But it's a good idea to take note of what you do and don't like so that you have a clear idea when it comes to making your plan.

OBSERVE YOUR GARDEN

Whatever time of year it is, I'm guessing that if you have a new garden and are reading this book, you don't want to wait. A lot of gardening books will tell you to wait a couple of seasons so that you can assess the garden, know what you have and then get started. This can be nothing short of torture if, like me, the day after moving in you have already made a plant list and ordered your greenhouse! There is, of course, a lot to be said for waiting. The art of waiting and observing is probably one of the most underrated in gardening. However, waiting and observing is usually more enjoyable when there is something to observe! If you are starting from nothing, waiting will just serve to irritate you. So, instead, we need to find ways of assessing our garden without waiting. Then we can get going on our plans as soon as we feel ready.

You can easily assess your garden and get started right away in the summer.

Observations of sunlight

Sunlight is one of the most important factors in any garden. It determines what you can plant where and may be the main limitation in your garden. Some gardens benefit from a lot of sun, and some not so much. In most

PLAN AND DESIGN YOUR GARDEN

Assess what's in your garden already

Look at what you have in your garden already. If you have some building materials, pots or ready-made borders or beds, ask yourself if you would like to incorporate them into your overall plan. Similarly, you might be able to use some of the infrastructure you already have to set you up for your future plans.

My garden when we first moved in.

GROW A NEW GARDEN

Different parts of the garden will be shaded at different times of the day, even in a south-facing garden like this one.

cases, however, the amount of sun reaching your garden will change throughout the year. If you have just moved into your house, you might not be aware of what the light levels are like over different parts of the year, so observation is important.

The sun moves in an arc, rising in the east and setting in the west. At higher latitudes, the sun will move much higher in the sky during the summer months than it does in the winter months. In my garden, I get very little sunshine until around March, when it starts to pour in like treacle, slowly bringing everything to life with it. In the winter, the sun is shaded out by an adjacent hill, my house, my garage, my neighbour's house and a row of mature birch trees at the end of the garden.

Identifying which way your garden faces might be something you did prior to moving in, in which case you might have a good idea of how much sunlight to expect. People who have gone to the effort of finding out which way their garden faces will understand that, generally speaking, south-facing gardens get the most sunshine and north-facing gardens get the least. To determine which direction your garden faces, you can either look on a map or use a compass. Your garden 'faces' in the

PLAN AND DESIGN YOUR GARDEN

A north-facing garden will be shaded by the house.

A south-facing garden will be quite sunny, with a little shade at the end if there is a fence, wall or hedgerow.

An east-facing garden will be sunniest on the right-hand side as you look from the house.

A west-facing garden will be sunniest on the left-hand side as you look from the house.

direction furthest from your house. For example, if the far end of your garden is in the south of your plot and your house in the north, you have a south-facing garden. It is also likely, by extension, that you have a north-facing front garden.

The direction of your garden matters a lot more if it is small rather than large. And the direction of your garden matters less than the actual light availability. For example, if you have a south-facing garden that is short and wide with a block of flats at the bottom, it's not going to have more sun than a long north-facing garden. Your house is usually the thing that is shading your garden, so, where the shadow of your house extends, sunlight could be limited. There might be other things that shade your garden, too, creating small pools of shade. This might sound complicated, but it actually just creates a more diverse range of planting conditions, opening up more options when it comes to planting.

Table 3.1. Garden orientations and features

ORIENTATION	FEATURES
North	A bit of a challenge if it is small.
	They are generally the shadiest of gardens, and they will benefit from shade-loving plants. If you are able to plant appropriately, north-facing gardens can be beautiful, verdant oases that on a warm summer's day will feel like a gift from heaven.
South	Generally gets the most sunshine.
	The main problem is trying to create shade: south-facing gardens can get very warm. Most common garden plants enjoy full sun or partial shade, so you will have plenty to choose from when it comes to planting. And if you are particularly keen on some of the more shade-loving varieties, you can create shade.
East	Experience shade in winter and sun in summer.
	They catch the morning light and by the evening the sun will most likely have disappeared behind your house. In an east-facing garden, you might want to plan a seating area at the end furthest from the house because this is where you are most likely to see the evening sun.
	As you are likely to experience more sun in the summer and quite shady conditions in the winter, using perennials can help make the most of the conditions. These are plants that go dormant during the winter, and then start to come back to life in around March. By using perennials, you can grow some plants that thrive in full sun and they will be none the wiser that they're in shade for six months of the year. Don't forget that even in an east-facing garden there will be pockets of sunshine and pockets of shade. Try to observe where these will be based on what is likely to be casting shade.
West	Similar to east-facing gardens, but with the evening sun as opposed to the morning sun. This means that if you like to sit out in the evening sun, you should put your seating area closer to your house.

PLAN AND DESIGN YOUR GARDEN

Don't forget, it isn't just houses that can shade gardens.

TO CALCULATE SHADE

It is almost impossible to accurately calculate where shade will be in your garden, as it depends on so many factors including your latitude, the local topography and, for many, the plants and buildings that surround your property. But you can make a reasonable estimate. Make sure to take into account trees that might not currently be in leaf. In the winter, these trees will cast little shade, but in the summer they could create an area of deep shade.

To estimate where the shade is likely to be in your garden, draw a plan of it and mark where north, south, east and west are. Shade the north side of any walls or fences, because these will cast shade wherever they are. The taller the wall or fence, the bigger your shaded area should be. You can then estimate shade being cast by trees or other structures in your garden, knowing that the sun will travel from east to west and will sit low in the sky in winter and higher in summer. The sun will always be slightly south if you are in the northern hemisphere. You can make separate plans for different times of the year or use different colours of shading.

Another way to gauge where the shade will be in your garden is to take a look at aerial photographs or old photographs. Google Maps might give you a pretty good idea of what you will be dealing with in a different season in your garden. You might have photographs from the purchase of your property that will also give you a clue. Similarly, you can ask your neighbours, who might be willing to give you some advice. If you can sneak a look over the fence into your neighbour's garden, you might be able to gauge roughly how much sunshine to expect. If there are plants already growing in your garden, you can try to identify them and see if they are sun- or shade-loving species. By piecing together these clues, you might be able to get a reasonable idea. Another idea that I've seen used with some success is to create a small model of your garden. This can be done simply with Lego bricks, if you like. You can then use a torch to replicate the sun and give yourself a pretty good idea of what to expect.

The most important thing is not to panic. You won't be able to calculate the shade completely accurately; there will always be surprises. So don't worry if it results in having to move a few plants around if they're not happy. It happens.

WORK WITH THE SEASONS

When starting a new garden or observing your current garden, it pays to be aware of the season you are in. Here are a few pointers on what the seasons could bring in terms of garden red flags and what you should be doing during these times.

Winter

Take the time in winter to assess your garden and make notes. Is there anything living in it through winter? Are there any areas that flood? Are there any particularly windy or exposed spots? Now is a great time to find out. Make sure you make notes, too, as these things will likely disappear from your mind or distort before you have a chance to act on them.

PLAN AND DESIGN YOUR GARDEN

Left, Observing your garden in winter can help you identify potential problem areas, such as this sunken area on my lawn that floods easily after heavy rain. *Right,* You can start sowing seeds in spring even before you have a design in mind, if you know what you would like to grow.

If you have brought plants from a previous garden, you can ensure they are looked after through the winter by keeping them in pots. If you already know where the plant will go, it can be planted out, but keep an eye out for waterlogging because that is generally the fastest way to kill most plants.

Spring

If possible, spring is the best time to jump into action. So, if you haven't already, you need to start coming up with a plan. Once you have the outline of a plan, you can start to create sections of your garden, planting as you go. This segmented approach will ensure you have something fun to observe and bring a pop of colour whilst you are working on other areas.

If you know some of the plants you would like in your garden, you can start cultivating them in spring, either on a sunny windowsill or in a greenhouse. Many wonderful plants can be grown from seed for a fraction of the cost. All you need is a bit of space, some seed compost and, of course, some seeds.

If you don't yet have a plan, you can accomplish the same fun and colours by planting some seasonal plants you like in containers.

Summer

Summer presents another opportunity to assess your garden, just like winter. Take note of any plants you wish to keep or remove. Plants are easier to assess and identify when they're in full leaf or flower. This season may bring pleasant surprises or reveal areas that need improvement in your garden.

Assess the soil in the summer. If it is cracking, you likely have clay soil, so watch out in winter when it might become waterlogged. If the soil is drying out and you need to water plants a lot, it is certainly lacking in organic matter and could be particularly sandy. There is never a bad time to mulch the garden, so you can do this right away if you see fit.

Autumn

Don't be alarmed if everything looks like it's dying in autumn. This is probably not its permanent state. Gardens can look particularly bedraggled this time of year, particularly after strong winds or heavy rain. But if you know where your beds will be, now is a great time to mulch your soils and start doing some groundwork so that your new beds are ready and waiting for you in spring.

CREATE YOUR PLAN

Now that you have an idea of how you want to use your space, as well as what style you want your garden to be, we can go about planning it.

You will need to start by making a map of your garden as it is. This doesn't have to include anything that you will be taking away but should include all your boundaries and anything that is staying. Unless you are a landscape architect or a garden designer, making a to-scale plan of your garden can be a bit of a challenge. I'm terrible at this, so I have devised some ways of cheating over the years. Here are some easy ways to draw a plan of your garden:

PLAN AND DESIGN YOUR GARDEN

Terrain map. Take a screenshot of your garden from something like Google Maps and trace around the perimeter, marking any structures, trees or shrubs that you want to keep. If you do this on a phone, tablet or computer, you can also make the plan larger, which is always useful.

A drone. You might want to check if you're allowed to use a drone in your area, but this can be an excellent way of getting a plan of your garden. You can then trace and use as you wish.

Your property plans. You may have had some plans in the documentation you received when you bought your home, which you can use to at least define your boundaries.

Once you have a rough plan, you will need to make some measurements to ensure that your plan is accurate. If you are only wanting to

Create a plan of your garden first, and then make copies.

make a few adjustments to your garden, you can get away with just measuring the bits that you are going to play with. Otherwise, measure your boundaries (you may need to do this in several different stages) and any existing structures you have, like patios, decks, beds and borders. Mark in any trees you want to keep, and don't worry if you need to move them around in later versions of your plan if they aren't quite in the right place.

If you want to do some hard landscaping in your garden and order paving slabs, wood, bricks, gravel or the like, you will need to be as precise as possible with your garden plan. Include as many measurements as you can so that your purchasing is accurate.

Once you have your plan indicating where your boundaries are and any structures or trees that are staying, I recommend making copies. Having a digital version may help, or you can just make a few photocopies. This duplication will enable you to make changes as you go along without creating too much of a mess on the original copy, and without having to redraw everything.

DRAW YOUR GARDEN DESIGN

It is at this point that we create a zoning diagram. This stage is where we look at where we want to do what in our garden. Where will you be hosting your dinner parties? Do you want to take advantage of the late evening sun, or place your dining area near the back door with easy access to the kitchen? Where will the kids be playing football, and will they be kicking a ball anywhere near your precious greenhouse? What kind of plants do you like, and where would they grow best? Do you love lush shades of green that would thrive in shade, or are you more into bright bursts of colour better suited for a spot in the sun?

You can then continue by drawing in any major structures you would like, including a patio, dining area, shed, greenhouse, deck, pond, vegetable patch, sun lounger or hammock, arbour, bench, or garden office.

These don't have to be things that you will be adding right away, but giving yourself a clear plan will help you to work around them and make sure that you don't waste time and money on an area that will end up with a shed on it. When placing elements on your plan, make

PLAN AND DESIGN YOUR GARDEN

This simple garden design uses light and shade and has areas specified for seating and repeat planting.

sure to take into account the sunlight, prevailing wind and other factors we discussed earlier in the chapter.

TIPS FOR DESIGNING A BEAUTIFUL GARDEN

These tips will help you tie in any element of your garden design. A few top tips from Kenny will help you to think about your space and how you want to design it.

Try to mimic the landscape

This might be easier said than done if you are living in a city and the last thing you want is to mimic the landscape, but the clever use of some vintage street furniture or concrete pots can help your garden work with, rather than against, its surroundings. Take cues from your surroundings to help your garden feel more natural and relaxed. Don't forget you can use dwarfing plants to create the effect of mini trees without taking up as much space.

Remember canopy planting
Don't forget about vertical space in your garden. If you plant everything at the same level, you are missing an opportunity to create a lush and interesting display. Use plants of varying heights underneath one another to create a more natural look.

Create space with curves
'In smaller spaces, the use of curving lines and circles can help make it bigger,' Kenny says. If you have a small, square garden, placing a round patio in the middle will give you some deep corners to work with. You can place larger shrubs or trees in these corners and then smaller shrubs and perennials in the narrower spaces. You can also use ground cover plants to soften the edges, to ensure that your borders are full and to stop your patio from being encroached on if it is very small.

Limit materials
'You want to create some uniformity in your garden,' says Kenny, 'so if you can, stick with the same materials throughout. For example, if you want to use pots, choose the same style of pot [–] all terracotta or all stone.' By using the same material throughout your garden, you can make it look like one whole garden. For example, the use of the same type of pavers for your patio and your path will help tie the two together. You can't create raised beds from pavers though, so choose something else for that, and make sure that you are consistent. Think of this choice the same way as choosing the furniture in your house; if it is all mismatched, the overall effect is a bit chaotic. Don't forget that you can use paint on wooden structures to blend them with one another and the surroundings.

Curves, such as this circular patio in Kenny's tiny garden, create space in a small garden.

GROW A NEW GARDEN

Left, Planning your planting can help save you money and give your garden a wow factor. *Right*, Limiting your colour palette helps make choosing plants easier and makes the planting look more coherent.

PLAN YOUR PLANTING

When you get to the point where your beds and borders are created, and your soil prep done, then you are ready to think about plants. In case you skipped the section on soil prep, I recommend you go back to Chapter 2, 'It all starts with soil' (page 31), and read it. But the general ideas are to add organic matter to the surface and try to keep soil

PLAN AND DESIGN YOUR GARDEN

disturbance to a minimum. If you've just moved a large amount of soil around, don't worry; just keep disturbance down from now on.

The temptation at this point is to head down to your local garden centre, accost the nearest clever-looking member of staff and ask them for some advice on what plants you should choose. Try to resist this temptation if you can, and instead come up with your own planting plan. This can be on the same plan as your overall garden plan, or you can choose to focus on specific areas, one by one. Just bear in mind that, unless you are able to section your garden into several different areas visually (using screens and so on), you will want to create some unity, so you will want to echo your planting throughout the garden. Don't panic; it's not as complicated as it sounds.

CHOOSE PLANTS AND A COLOUR SCHEME

You know what style of garden you would like by now (or at least you have a rough idea), so now it's time to choose your colour scheme. Once again, garden designer and close friend Kenny Raybould is here with some advice. 'Limit your colour palette,' Kenny begins. 'We often want to buy everything, but for the best look you should choose three or four colours and work with them.' He gives the example of pink, purple, white and green. 'Never neglect the power of green. Use green in several different shades to create texture and shape in your garden. When working with pink, choose two or three different shades of pink and repeat them throughout the garden, using several different flowers to represent the colour. Try to use different shapes of flowers, too, to create more interest. Try not to go straight to the garden centre and buy everything in flower; write down what you want first and research it, so you know you're getting the right thing for the conditions in your garden. Make sure it will be getting the right amount of light, the correct soil and whether it will be able to cope with other pressures such as the prevailing wind.'

To make this process easier for myself, I looked up plants that work well in the conditions of my specific garden: partial shade or full sun, and that enjoy clay soil. Most lists online or in books come with pictures, which is helpful if you aren't a walking plant encyclopaedia like Kenny. From the pictures, you can get a reasonable idea of what colour foliage or flower to expect. I made a list of plants in various colours that I knew were suited to the conditions in my garden. I noted how big these plants get, as well; height and spread are important. Once I had a list, I created a mood board using images of these plants to give myself an idea of which ones will work together. I also took inspiration from images in books and online, and from gardens I had seen in real life. If there was a particular combination I enjoyed, I would find out what those plants were and research them. They would either be added to my list or discarded, depending on whether they would be suited to my wet, heavy-clay, east-facing garden.

Kenny recommended that I draw out how I want my bed to look: 'Understand how far your plant will spread, and then give it about two-thirds of that space; you really want plants to thrive, but you want them to grow into each other, too.' This should give you an idea of how many plants to buy. 'To create uniformity, it's really important to repeat planting.' Don't just buy one plant; buy three, five, seven, or more of the same plant to create a continuous theme throughout your garden. Plant them in clumps if they're small and repeat them somewhere else.'

Once you have drawn out roughly how you want your bed to look, then you can start buying and ordering plants. Don't forget to take a look at 'Choosing healthy plants' in Chapter 7, page 179.

TIPS FOR GREAT-LOOKING GARDEN BORDERS AND BEDS

Here we are, we've arrived at the point of filling your beds. Hurrah! This is the fun part. After this, you can begin to enjoy the plants you are growing.

Once I had carefully made my planting plan and then bought the plants, I raised them in my greenhouse for a little longer. I bought most of the plants for my garden as plugs or grew them from seed. To ensure that they weren't eaten immediately, I wanted to get them to a better size before planting them out. This was in early spring. By late spring, growth was so vigorous I was no longer worried about them being eaten, because they would probably be able to outgrow the munching from my local molluscs (slugs and snails).

Planting in clumps can help your garden look more natural.

GROW A NEW GARDEN

Top, Filling your beds is definitely the fun part! *Bottom, Lysimachia* spp. and fleabane spilled from my beds in year one.

At this point, I had entirely lost my planting plan. It had probably been flipped over, made into a beautiful work of art by one of my children and then stuck to the fridge, but I couldn't find it. So instead, I carted all the plants out of my greenhouse, in their pots and laid them out on my borders. I fussed with them until they looked about right. And as it would happen, I still continue to fuss with them. In fact, as I write this, I am continually glancing out of my office window at a grass I have just decided needs to come forward by a foot.

As I was on my hands and knees moving plant pots around, I gave Kenny a call. Luckily, my dear friend was, as usual, brimming with useful information. My beds are parallel to my house, and wide, so I don't want anything too tall in them. I have a cherry tree at one end, which helps give me some privacy from the neighbours, and a birch tree that ties in with the three birches I have against the back fence and softens the view of the office from my kitchen window. Kenny advised me to put some of the taller plants in the middle of the beds, with shorter ones at the sides. Where the beds drop down to the grass, I put some trailing plants. The basic principle is called 'killer, filler, thriller and spiller'. The killers are the few plants that steal the show. In my garden are *Salvia* 'Amistad' and *Dahlia* 'Bishop of Llandaff'. The fillers were asters (which became thrillers later in the year!), yarrow, grasses and other foliage plants, veronica and kale. The spillers were *Erigeron* spp. and *Lysimachia* spp.

TIPS ON PLANTING UP BORDERS

Kenny seemed to want to educate me to the level of a professional garden designer, so I listened to him intently as he gave out more advice on planting up borders.

Repeat your plantings

Repeat planting or repeating the same advice? This tip is being repeated to you just as it has been repeated to me, numerous times. This is because it's really important for creating unity, reinforcing a theme and providing a natural look for your garden. 'Just as we want to repeat materials and limit ourselves on that, we should also limit and repeat with plants', says Kenny. 'You can also do this with communities of plants. If you have a few plants that look nice together, put them together in a few other places, too, to create an ongoing pattern.'

Blend colours and textures

Kenny and other masterful garden designers use plants to paint their gardens. Just as an oil painter would blend colours to soften edges, you can do the same in your garden to create a more natural, harmonious look. Similarly, you can choose to clearly demarcate your colours and not blend at all, like in formal gardens. Kenny, whose style, like his personality, is creative and a little eccentric, advocates blending. 'For example', he says, 'you can have a group of one plant in the middle of a border, and then a couple trailing out from either side. When mixed in with other plants, this will look more natural.'

Green-up your walls and fences

'If you have walls or fences in your garden, grow something in front of or over them. Plant your climbers at the back, against the wall or fence. Allowing them to grow up and cover your walls or fences will help to dissolve them into the garden, and make the space look bigger', says

GROW A NEW GARDEN

Greening up your fences and walls can help blend your garden into its surroundings and make it feel bigger.

PLAN AND DESIGN YOUR GARDEN

Left, My garden in June of its first year. *Right,* The same view in June the following year.

Kenny. You can also use trees to the same effect, but they will take more space out of your garden and may cast shade on it where you don't want it. Choose what is appropriate for you and your garden, and make sure it ties in with the rest of your planting scheme.

Take growth into account

'By making yourself aware of a plant's growth rate, you can give it the space that it needs. Plants that are overcrowded won't do as well', Kenny says. 'Bear in mind that some plants will grow differently according to their environment, too. Some plants will surprise you with how big or small they get, so don't worry about making changes.'

GROW A NEW GARDEN

I continue to edit my own masterpiece.

Set up planting communities

'Planting communities are groups of plants that coexist in nature, or that enjoy growing in the same kinds of condition', Kenny explains, saying that this term is commonly used but the meaning often overlooked. 'We have a habit of buying 20 different plants because we like them and not considering if they will all be happy growing together in the same conditions. By curating planting communities, we will have more success with creating a natural look and growing healthy plants.'

Know the plant

This tip is from me. Make sure you understand what to expect from the plant by the time you first put it in the ground – not only its size but also its behaviour. Some plants will spread at will and take over your bed. Some will not cope well with being too close to larger plants. Some plants don't like to have wet roots over winter, and if your garden gets wet you might lose them. You will learn more about the plants in your garden as you progress with your gardening journey, and there will always be surprises.

Kenny says to 'enjoy learning as you go. A garden is never finished, and you will always enjoy editing your masterpiece. So enjoy it as much as you can.'

Different flower shapes

Insects differ in shape, size and mouthparts. If you want to attract insect pollinators to your garden, offering flowers in a range of shapes

PLAN AND DESIGN YOUR GARDEN

A note on invasives

Some plants are invasive, so bear in mind that a few of plants I have mentioned above, whilst not invasive in my garden, might be invasive in yours. Invasiveness is all about context. In Britain, one of our worst invasive plants is Japanese knotweed (*Reynoutria japonica*). Its presence in your garden can invalidate your home insurance and make it impossible to sell your house. However, in Japan it is not invasive because it has several natural controls to its growth.

One would hope that, when buying a plant from a local nursery or garden centre, you wouldn't be introducing an invasive plant. But importing seeds, taking cuttings of something you like the look of, or buying plants from spurious sources can lead to this kind of mistake. Always check before you buy, particularly if it is a species you see growing in the local area a lot.

and sizes is a good thing. Many flowers nowadays are bred into pom-pom shapes, lacking the centre of more traditional flowers and having instead a cluster of petals. This is often seen on roses, dahlias, cherry blossoms and other similar flowers. If you want to attract pollinators to your garden, and I highly recommend that you do, these flowers aren't ideal. They make access to the nectar difficult for most pollinators, so try to find flowers that have an obvious centre, even if it is tiny tubular flowers in clusters, such as *Salvias*.

REMEMBER WHAT YOU WANT

There are lots of other ways that we can create enjoyable spaces in our gardens. Some years ago, I had the pleasure of judging a local gardening competition that was open only to people in council-provided accommodations. Their budgets were next to nothing, but their imaginations were utterly limitless. One lady created a spectacular tiki bar in her garden, complete with numerous circular tables and chairs

PLAN AND DESIGN YOUR GARDEN

A note on native plants

Native plants are amazing; they will be well suited to your climate and will probably grow really well. There are also a lot of benefits to nature, with some species of insect having particular tastes and preferring only certain native plants. There are a lot of benefits to growing native plants but don't be afraid to grow non-natives, too, even in a wildlife garden. There is a lot of evidence showing that most pollinators are generalists, and the presence of nectar produced by any flowers is a good thing, whether it is from a native plant or not.

occupied mostly by stuffed monkeys. Another decorated an entire hedgerow with what looked like a lifetime of little trinkets, from teacups to crystal angels. Another lady had created a Japanese garden using blue stones in place of water and a shower curtain that looked like a bamboo forest hung against the back fence. There were also a good number of gardens that were just as interesting, cosy and fun but in perhaps more mainstream ways! The thing I loved about these gardens was that the tenants used what they had to create places that were fascinatingly unique and quirky. And, mostly, places that they themselves loved. So, if monkey tiki bars or teacups in bushes are your thing, why not go for it? It's your space; you make the rules. If you're a bit more 'down the line', the same applies: you make the rules. If someone says you can't have pink and orange together, but you love them both and simply can't choose between them, it doesn't matter. Do what makes you happy.

When it came to my own garden, I did throw out the rule book to some extent. Over my years of gardening, I have fallen deeply in love with a few plants, including *Rudbeckia* 'Goldsturm' with its vivid, black-centred yellow flowers, and *Thalictrum delavayi* (Chinese

Using different-shaped flowers, like these *Calibrachoas* and *Salvias*, looks great and caters to lots of different insects in the garden.

meadow-rue) with its frothy clouds of insect-like purple and white flowers. There is also an open-centred white rose called *Rosa* 'Rambling Rector' that I originally bought when my grandfather died because I thought he would appreciate the name, and I hoped it would remind me of him. I have since planted it in every garden I've had, even in rental properties. Flowers in white, yellow and lilac. A garden designer's worst nightmare! But these are the things that make me happy, so I want them in my garden, and I'm not prepared to be told otherwise. They're like old friends. So, my garden ended up being a mix of purple, yellow, white, red, blue and even orange. The orange was actually by accident, but it set off the other colours beautifully, so I ended up adding more.

Don't be afraid to make changes

At one point during my first year in my most recent garden, I planted hundreds of plug plants into the beds. It takes quite a lot of imagination and knowledge of plants to know how something will look when you're only planting a stem with a few leaves, so don't be afraid or ashamed to make changes later if you need to. Perennials in particular make up the backbone of many gardens, and they really don't mind being moved. Most will carry on as if nothing ever happened.

When moving a plant, make sure that you dig up as many of its roots as possible, trying to avoid damaging them. It is far easier to move a plant during winter than during summer because it won't dry out so quickly whilst its roots reestablish. So if you can wait, do. But with some careful handling of the roots and plenty of water when they are put into their new position, most plants will do just fine being moved in the summer. I'll let you decide whether you want to take the risk or not. You should always move a plant as soon as you can after planting. If it's a year after you planted it and it's not thriving or its position isn't working visually, move it now rather than another year down the line.

If you are moving a plant during winter, make sure the ground is not frozen. After digging up as many of the roots as possible, make a hole that is a bit bigger than the root ball, add some organic matter into the bottom of the hole and gently place the plant in the hole. Backfill and water the plant in thoroughly. Transplanting in winter allows a plant to concentrate on root growth before it is required to make flowers and new leaves in spring.

PLAN AND DESIGN YOUR GARDEN

If you have taken over a new garden with some large plants, and you're not sure how old they are, be aware that moving them might kill them. Plants that are well established (i.e. have been in the ground for several years) will not thank you for moving them. That being said, you can get away with moving some plants. However, if it's something you particularly like, I recommend trying to incorporate its current position into your design rather than trying to move it. If it's in the way, making changes, such as pruning or tying it back, can help it to fit in a little better.

CHAPTER 4

Materials in your garden

Choosing the materials to use in your garden often comes down to preference and budget. If there is some infrastructure in your garden already that you intend to keep, your materials might be largely dictated to you. There are many considerations that need to be made when considering what materials you will use. The longevity, cost and upkeep are just a few of the main points. If your budget is low, you may be looking for reclaimed or second-hand materials, in which case the availability of items will come into play. If your budget is a little larger, you may want to consider getting something that will last a long time and require less maintenance. Ultimately, there are pros and cons to most things, and it's about choosing something that matches your needs, budget and taste. If you are unsure what to use, we will take a look at a few options.

CONTAINERS, POTS, RAISED BEDS OR IN THE GROUND

When it comes to designing the actual growing space in your garden, there are several options. Many people believe that raised beds represent the highest form of garden bed, but that's not necessarily the case.

The vegetable patch I created from scavenged materials: one side is concrete blocks covered in render and the other is wood, because I couldn't find enough of either to complete the job.

GROW A NEW GARDEN

So, let's look at some of the options, along with the pros and cons of each.

Raised beds

Raised beds have the obvious advantage of being a little higher than the ground. Some, like the ones in my garden, are really high and offer a definite advantage when it comes to not having to lean over. I can reach the entire bed without having to bend at the waist. This is helpful to me now in my late 30s, but it would be priceless if I were elderly. On the other hand, if you want to garden with your children, high raised beds aren't good for providing access. If you are building raised beds from wood, bear in mind that, when wood is constantly in contact with soil, it will rot quite quickly unless you buy good-quality, thick, pressure-treated wood.

You don't need to do one or the other, but have a clear idea of what you will use where.

Some other pros of using raised beds include:

- **Aesthetics.** There's no denying that it's satisfying to have a clear demarcation between path, lawn and bed.
- **More light.** By being raised and filled to the top, raised beds can let your plants have more access to light. This usually means that you can pack more plants into them.
- **Better drainage.** By raising the grounds up, you can often achieve better drainage within the bed, useful in areas where drainage is a problem.

On the other hand, raised-bed cons include:

- **Hiding places.** They provide an opportunity for small insects, slugs and snails to hide. The edges of raised beds are particularly

MATERIALS IN YOUR GARDEN

Raised beds can make things a lot easier for people with mobility or back problems.

good for pests to hide and overwinter. By creating raised beds, you have essentially created a habitat for them.

Cost. They can be quite costly to make. If you can find the materials cheaply, this can be negated.

Mowing. Mowing up to or between raised beds can be difficult, so if you have grass next to a raised bed consider how you will keep the edges tidy.

Containers

Containers, in my opinion, are hugely underrated. Perhaps the best thing about growing in containers is that they are so versatile. You can change around your garden any time you like, and, if you are in temporary accommodation such as a rental property, you can take the whole garden with you when you leave. There are, of course, some challenges with container gardening.

Pros of growing in containers:

- **Hugely versatile.** You can move things around easily, and there are lots of shapes and sizes to use.
- **Tiering.** You can use them to mitigate slopes in your garden and create good-looking tiers.
- **More light.** Often, plants have more access to light and can be planted closer together.
- **Different soils.** You can grow plants with different soil needs next to one another in different pots.
- **Moveable.** The whole garden can come with you if you move.

Container-garden cons:

- **Cost.** The best container gardens are curated over many years, for the simple fact that buying lots of pots is quite expensive. You

GROW A NEW GARDEN

Pots or containers are a great choice if you want to be able to set up a garden quickly and move things around.

can pick them up second-hand easily, and you can start small and build your collection.

Fertilisers. You can't rely on the processes in the soil when you are growing in containers. Potting compost will need to be changed from time to time – depending on the plant – and you will need to stick to a fertiliser regimen (again, depending on the plant).

Size limit. It's difficult to grow larger plants. Whilst they may grow well, keeping them safe from high winds can be a problem. Small trees in pots often fall over during high winds and break their pots.

Straight in the ground

Good old-fashioned borders and beds in the ground are no-nonsense, cost-effective and always a joy. You can grow plants of any size; they have access to the soil, which means they will be fed and have better access to water. Beds and borders can also be customised and changed to fit your needs. Want more flowers in your garden? Widen your border! Want more lawn? Make your border smaller. Want to add interest? Create some island borders in your lawn. The possibilities are endless! Of course, the best option of all is to have a mix of these elements, whatever suits you.

Pros of planting straight in the ground:

Cost-effective. You can use as much or as little of your garden as you like for flower beds and the cost stays the same.

MATERIALS IN YOUR GARDEN

Flexible. You can widen a bed or make it narrower, change the shape; all it takes is to move a few things around and open up the ground.

Plant health. Plants have direct access to the soil to get hold of nutrients and water. You can grow any size of plant, depending on the space you have available.

On the flip side, if you plant directly into the ground you will have to maintain the edges to keep them looking neat and to stop two parts of the garden from merging – unless that is a look you like, which is also totally OK!

BUDGET-FRIENDLY WAYS TO FILL RAISED BEDS

Filling raised beds can be a particularly costly endeavour if your intention is to fill the whole thing with compost. If you are making raised beds in the ground, the chances are you will have some spare soil, so that should absolutely go into your raised bed. There are also a few other options to fill your raised bed, though it's worth making sure that no less than the top few centimetres (or 1in) is compost or at least something that you can plant into. Generally speaking, put larger fill at the bottom and finer fill at the top. Don't use anything too large if your bed is quite small. Try to use what you already have as a first port of call. Here are some examples of the materials you can use to fill your beds:

Wood. Woodchips, logs, sawdust, branches and leaves – anything woody will break down slowly and help to support fungal growth in the soil. This is particularly good when you are growing perennials, shrubs and trees, but it's not so good for annuals and vegetables because they tend to rely more on a relationship with bacteria than with fungi. Therefore, creating a fungi-rich soil isn't necessarily what you want, but it certainly won't do them any harm. If the wood is too close to the surface, it may interfere with your ability to plant into the bed.

Rock, stone or rubble. You may have rocks or stones at your disposal, and they are fine to use, providing that they are mixed in with something finer. The aim here is to create height and fill your bed, without creating a barrier between the planting surface and the soil beneath. Larger stones work in larger beds, but make sure they won't get in the way of your plants. You may want to consider what you want to plant before using any stones. For example, you don't want stones in a shallow bed intended to grow carrots. This will interfere with their growth and produce some interesting-shaped crops. However, if you are planting some alpines, you can use almost entirely stones with little soil. Please be careful if you are adding rubble. Ensure that it doesn't contain any contaminants, such as asbestos, that could seriously harm you and your garden.

Sand. Sand can be quite a cost-effective way to add bulk to your soil and compost without changing the texture too much. Sand is known to add drainage, but without the organic glue to help soil structure stay together it is not much use. However, if you have some sand going spare and you need to fill up a raised bed, go for it.

Soil from elsewhere. If you are planning some raised beds, it's often prudent to think about how you will fill them before making them. If you have other projects on the go, or are likely to do something that will result in you needing to get rid of some soil, try to pair the two projects together. This method is simple and easy and does a good job. You may need to pair it with some other materials to top up.

Green waste. Green waste is a type of compost that is often available for free from your local recycling centre. Sometimes the quality isn't great, but your local green waste might be pretty good, so it's worth asking around to see if anyone has tried it, or taking some for yourself and giving it a go. This can be a cheap and relatively easy way to fill a raised bed, but you should pair it with one of the other things on this list or you will find the level sinks down very fast!

HARD LANDSCAPING

Hard landscaping involves creating paths, patios, decks, steps, walls, fences, drains and other such features. This part can be genuinely costly, but doing it yourself, if you are confident and physically able, can be an excellent way to save a huge amount of money. Before embarking on any hard landscaping, however, you should do some

MATERIALS IN YOUR GARDEN

You can get creative with how you fill your raised beds, but beware of creating habitat for plant-eating pests.

research into best practice on whatever it is you want to build, and, of course, make a plan. If you decide to do any hard landscaping, here are a few basic tips to set you in the right direction.

Choosing materials

If you are planning to embark on a DIY job in your garden, remember that the materials you choose to use can have a huge effect on how easy it is to complete your design. Some materials will be fairly obvious choices. For example, if you want a patio, you will have to use slabs, and if you want a deck, you will have to use wood or composite plastic. For other projects, however, here are some tips gained from a lifetime of trial and error!

Wood. Easy to use, quite forgiving and generally requires no specialist tools. Ensure that you choose a thick, pressure-treated wood so that

It's important to get the hard landscaping right, so consider this element carefully before spending any money.

it will last. Thin wood might seem like a cost-saving exercise, but if you fill it with soil, or it is exposed to the elements, you will end up having to replace it regularly. Timber sleepers, such as railway sleepers, cost more but will last a lot longer and cost you less in the long term. They are available in a range of different materials, some harder and more durable than others.

Stone or brick. A challenge to work with if you don't have the skills or the tools. I recommend calling in the professionals or doing some project-specific research before embarking on anything using stone or bricks, especially as the materials are more expensive.

Flooring materials. For a grass-free area, such as a path or hard-standing. It can be difficult to choose what material to work with if you want to create these grass-free areas. Generally, we choose grass-free because we want something that is low-maintenance, so choosing the right materials, and understanding how they will work over time, is essential.

MATERIALS IN YOUR GARDEN

A note on weeds in paths

It's a good idea to clear weeds away before laying paths, patios or decks, but seeds will remain. Seeds can also drift in and land on top, so, whilst a good-quality weed membrane can help, it cannot solve the problem entirely, and regular maintenance will be needed to keep them looking good (see Chapter 8, 'Managing pests and weeds' on page 185).

Even concrete slabs aren't enough to keep weeds at bay, so be prepared for some maintenance.

- **Slabs.** Relatively easy to lay but requires the correct foundation. If you don't put down the correct foundation, you risk the slabs being uneven and potentially cracking. Slabs that are close together will work well for weed suppression but will need some weeding between the cracks occasionally. Other than cleaning, they require little to no maintenance.
- **Gravel, shingle, stone or slate chippings.** Easy to lay and relatively cost-effective, but difficult to maintain. If you have overhanging deciduous trees, you will need a leaf-blower to stop the area from becoming covered with leaves in autumn. A heavy-duty weed suppressant mat is a good idea, but bear in mind that weed seeds can fall and germinate in the rocks, so weeding will still need to be done occasionally to maintain their look.
- **Bark or woodchip.** A more natural look. Whilst a great choice, as an organic product, they will break down over time. Woodchip lasts between one and three years and bark chip between two and five years, depending on the use and the conditions. They are easy to top up when you need more, but bear in mind the future cost and maintenance. Both are good at suppressing weeds, but as with stone and shingle, you will still get weeds, and it will need weeding regularly.

CHAPTER 5

Creating a nice space

For most of us, privacy will be the main consideration when it comes to creating a relaxing garden environment. But there are other factors that, when correctly managed, can make your garden a much more enjoyable place to be. The nicer your garden environment, the more inclined you will be to go outside and enjoy it!

Creating a nice outside space will make you spend more time in it.

GROW A NEW GARDEN

A water feature, even a small one, can make your garden feel more tranquil.

Depending on where you are and the particular circumstances of your garden, you will have different requirements. Reducing noise, wind and intense summer heat might all help to make your garden more comfortable.

One of the best examples I have seen of this came in the form of a 2023 Chelsea Show garden. The Memoria & GreenAcres Transcendence Garden, created by Gavin McWilliam and Andrew Wilson, consisted of a gentle colour scheme – white with a subtle hint of yellow and purple – along with a stone path and, at the end, a large stone wall with an angled top hanging over part of the garden. Through the middle of the wall ran a waterfall. It is entirely unachievable unless you are extremely wealthy, have no planning restrictions and don't mind running a waterfall off your power supply. But, when you stood next to the waterfall, it created an echo chamber, with the noise of the falling water bouncing off the stone. This had the effect of blocking out almost all the surrounding noise. Suddenly, you were transported away from the hustle and bustle of Chelsea Flower Show to a tranquil waterscape. It was very impressive. The perfect antidote to excessive background noise. I'm not suggesting that we all rush out to build 6-metre-tall walls in our gardens, but there are certainly some lessons to be learned from it.

A water feature is an excellent way to add a soothing feel to your garden. Having moving water in the garden can vastly improve the feeling of zen you get from being in the space. Water and soft hues are by far the best way to create a relaxing feel in your garden.

Another feature to consider for a soothing feel to your garden is soft, moving plants. These usually come in the form of ornamental grasses. They move on the slightest of breezes, creating movement and texture in your garden.

CREATING A NICE SPACE

Light is another nice way to improve the feel of your garden. You can make a garden feel quite cosy by adding some festoon lighting, uplighters, downlighters or path lights, especially in places with seating.

CREATING PRIVACY

I remember clearly my grandparents' garden when I was growing up. They lived in a 1930s house, and the garden was the width of the house plus the driveway and was divided into two sections. Closest to the house was an ornamental garden, complete with dahlias, busy lizzies (*Impatiens*) and a goldfish pond, like all self-respecting gardens in the eighties. There was a fence halfway down with a gate in the middle, and past the gate was the vegetable patch. I will wax lyrical about the

A private space can instantly feel more appealing than one that is overlooked.

vegetable patch that started my love of gardening later in the book, but for now I'd like to turn my attention to the fences. On one side, the neighbours were quite rude and unfriendly. They were given a 6-foot fence and never seen or heard whilst we were enjoying the garden. On the other side, the neighbours were friendly. In fact, the neighbours for several houses in a row were all friendly. They all had low, roughly 4-foot fences dividing their gardens. And so, whilst my grandmother was outside tending to her azaleas, her neighbours, all in their sun hats, were chatting over the garden fence. When I think about this now, it feels almost utopian. Most of us are not lucky enough to live next to people who we find so agreeable that we are willing to share our outdoor time with them, passing home-made biscuits over the fence and sharing recipes. For most of us, we want privacy in our gardens. Whether you have rude or over-friendly neighbours, you live next to a bus stop or you like sunbathing in the buff, there are plenty of good reasons to want privacy in your garden. After all, if we wanted to be around lots of other people whilst outside, we would probably go to the local park. Of course, it isn't always possible to put up a 6-foot fence and block everyone out. There are restrictions on boundaries; if you are on a slope, a fence might not do much; and sometimes budgets stand in the way of us doing what we want. Happily, when it comes to creating privacy in your garden, there are many different options.

The trees at the back of my garden mask the building behind us during summer, but in winter the privacy is lost (although we use the garden less then anyway).

Trees

Trees are the single best way to create the nice feeling of privacy in your garden whilst also bolstering biodiversity and using vertical space. They will rustle in the breeze, bring chirping birds and overall do a great job of blocking sight and sound. However, they do need to be used in the right places, and you need to select the right kind of tree. Consider that trees can cast shade, drop leaves and also create dry patches under their canopies that may need irrigating.

A single tree is obviously not a physical barrier to much, but it can greatly help to create privacy in certain areas of your garden. When you are sitting in your favourite spot in the garden and you have just one line of sight that needs interrupting, say from a neighbour's window, a tree is a great option. To find out where best to plant your tree, all you need to do is sit where you would prefer not to be overlooked and mark a line between you and the overlook. The tree can go anywhere along that line. Bear in mind that if trees are planted too close to a border your neighbours can choose to give them a haircut, so if you want a nice, healthy-looking tree that grows to an old age and a good height, don't put it right against a border or fence; give it a little more space. Bear in mind also that trees cast shade, so you must be prepared to deal with this as a consequence, whether the shade will be falling on your side of the fence or your neighbour's.

If, like at my current house, you are completely overlooked in your garden and there is nothing a tall fence could possibly do about it, a row of trees might be your only option. We have an apartment block that towers up behind our garden, which not only looks awful but also creates the feeling of being in some kind of amphitheatre. Fortunately, there are three mature birch trees planted along the back fence, and when they come into leaf in spring the entire block is obscured from our view. Our garden becomes a tranquil and secluded oasis once again! So, a line of trees can work well to break up a view and to provide some privacy.

Evergreen trees are a good option for creating year-round privacy, and they are also quite fast-growing. But they can dominate smaller gardens and create too much shade in the winter months. Deciduous trees will lose their leaves in winter, but if you mainly use your outside space in summer this shouldn't be too much of a problem. They will allow some light into your garden during the winter, and this is particularly important if the tree is likely to cast shade on your house. Whilst this shade may be welcome in the hot summer months, during the winter most of us in more northerly climates need as much light as we can get.

GROW A NEW GARDEN

Trees can be costly, so make sure to do your homework about what trees will enjoy the conditions of your garden. Ensure you know what type of soil you have, how exposed your garden is and the lowest annual temperatures that are expected and, if in doubt, ask for advice at your local tree nursery or garden centre. If you're planning to buy a tree but can't spring for one that is already fairly large, you will want to purchase something fast-growing. Even fast-growing trees will take several years to create decent privacy. Consider planting it in a raised bed or large container to add height, or opt for a well-supported climbing plant instead.

Hedgerows

Hedging plants are an excellent choice and have been used for centuries to create privacy and reduce noise in gardens. Hedges also create shade, and they even filter pollutants from the air. They are great for creating habitats for small animals, birds and insects, too – provided you choose the right species to create the hedge, such as native, berry-bearing plants. By choosing a mixture of trees, shrubs and climbing plants, you can add visual interest to your hedge and make it more appealing to wildlife. You may find that some species thrive more in your garden than others, and choosing mixed hedging will give you a better chance of finding something that will thrive. Often, hedging will come in packs that premix species tailored to particular needs, such as thick

Adding a tree to a large raised bed can give it height to help obscure overlooking windows, whilst the ageing fence extensions will support climbers.

CREATING A NICE SPACE

hedges, wildlife-friendly hedges and tall hedges. A native, mixed-species hedgerow will only really work in a larger garden or as just a section of a small garden because they can be quite space-consuming.

Bear in mind that hedges will cast shade onto your garden if they are anywhere other than the north edge of the garden. Any solid boundary will do this, but hedges will need maintenance so that they don't get too tall and boisterous. Make sure that you have at least some shears and a pair of loppers to do this. If the hedge is large, opt for a hedge trimmer and pruning saw to make your life easier. Either way, make sure to prune outside of the nesting season, because in some places it is illegal to disturb nesting birds. Of course, looking after your birds is always best as they are excellent at pest control.

Hedgerows can create privacy and shade, cool the garden and even filter noise and pollution.

Fence extensions

Fence extensions slot over the existing fence posts and allow you to add height to a fence using trellis panels. As fence height is often limited by law or local regulations, you will need to check how tall a fence you can construct before putting one up. If a tall fence is not legal, or practical, you can use fence extensions to help create a little more privacy. This works particularly well when paired with a climbing plant. If you want to create privacy in the summer and light in the winter, choose a deciduous plant.

Arbours

If you want to create a cosy and private area in your garden that provides shelter from above and the sides, an arbour is fabulous choice. Consider that creating an arbour may cast shade on areas next to it. For instance, I would love an arbour on my patio, but this would cast far

A well-placed arbour can create privacy and shade.

too much shade on my living room, which currently gets much sun. There are plenty of premade arbours that are great choices, or you can opt to build your own if you have the skills.

BEST PLANTS FOR CREATING PRIVACY NATURALLY

When it comes to solving a problem, I almost always turn to a plant as my first port of call. Feeling bloated? Dandelion. Need to calm down? Chamomile. Can't sleep? Lavender. Want to build a table? Oak tree. Need to fill an awkward space in your home? Houseplant! When it comes to creating privacy, plants are a great option and can be quite cost-effective, too. We discussed trees and hedgerows in the previous section, and these are obvious and great choices for creating privacy naturally. Here are some other types of plant that you might want to consider.

CREATING A NICE SPACE

Climbing plants

Climbers are an excellent way of creating privacy relatively quickly. They work particularly well when allowed to climb up a fence or over a trellis. Many will also do so without taking up much space in your garden. There are a lot of different climbing plants to choose from, so you should be able to find one that grows well in the conditions of your garden. The only thing that you must consider is that they will need something to climb up, so consider how you will support them. If you have a low fence or wall, attaching some trellis to the top can be a great way of creating privacy using climbing plants that is unlikely to upset your neighbours, or break any laws or regulations (though if you are extending the height of a boundary, it is always worth checking what the regulations are in your neighbourhood). Plants such as ivy, clematis, rambling roses, jasmine and Virginia creeper are all excellent choices for privacy, depending on their suitability in your garden.

Bamboo screens (Caution!)

Bamboo is a popular option for creating privacy in the garden because it is fast-growing and quite attractive. However, bamboo also spreads really fast and can easily become a problem even in larger gardens. When grown in a container, it can be stopped from spreading but will be prone to catching the wind and falling over. When planted too close to your house or allowed to spread close to your house, you could even be looking

Some climbing plants are fast-growing and can give privacy quickly; just watch out for invasives.

at structural damage or a reduction in the resale value of your property, so be very careful with bamboo. Avoid it altogether if you are not sure you'd be able to manage it. As we have already seen, there are plenty of other options.

Ornamental grass

I love using ornamental grasses as they are quite subtle, and don't look like you are purposely trying to shut out the neighbours. Some are also quite delicate, acting more like a lace curtain than a blackout blind. The right grasses, in the right places, can create a natural-looking privacy barrier and add some movement to your garden, which is almost as relaxing as not having your neighbour peering over the fence! Try to opt for large grasses, such as pampas, but be aware that pampas, silvergrass and *Stipa gigantea* can be invasive. Again, make sure they are appropriate for your garden conditions and will not be invasive.

REDUCING NOISE

Reducing noise in the garden is similar to creating privacy. Noise can be oppressive in a garden, particularly if it is constant. If you live near a main road, you will know this. Noisy neighbours, construction, road traffic, schools, playgrounds and air traffic can all make your garden a less pleasant place to be. Creating a nice soundscape in your garden can significantly improve your experience of your garden.

Blocking it out

The first thing to consider when trying to reduce noise in your garden is where the noise is coming from. Sometimes, creating a boundary such as a fence, wall or hedgerow can help to block out some low-level noise, such as traffic.

Noise from plants

Another tactic for reducing the effect of noise is to create white noise in your garden. Look for leaves that rustle in the slightest breeze, like cottonwood or poplar (*Populus* spp.), or plants that make satisfying

Ornamental grasses are a soft touch when it comes to creating privacy.

GROW A NEW GARDEN

Plants themselves, like this honesty, can create soothing rustling noises.

rustling sounds like honesty (*Lunaria annua*), quaking grass (*Briza maxima*) and rattlesnake master (*Eryngium yuccifolium*).

Soothing noises

Soothing noise in the garden is almost always created with flowing water. If you have a small water feature that makes a pleasant trickling or babbling sound, your ear will be drawn towards this instead of the less pleasing noises outside your garden. Other soothing noises to block out other noises include wind and wood chimes, and encouraging birds, who will bring chirping, chattering and a range of songs.

An echo chamber

If you have a lot of noise, or the noise is coming from above, such as air traffic, you might want to take a similar approach as with privacy and just focus on one particular area in your garden. With this method, you can enclose an area on three sides and put a roof on it, creating a small garden room. This works well as somewhere to relax, eat or entertain. You can double down on this by creating some soothing noises within or near your enclosed area. One of the best examples I have seen of this was a small gazebo with some seating and a small pond with a water feature. It was relaxing to sit next to the water, and, within the walls of the gazebo, the sound of the small water feature trickling into the pond was all you could hear. I could have stayed there all day.

CREATING SHADE

Much like creating shelter and reducing noise, shade is an important factor in making your garden more habitable, especially if you have a small south-facing garden. Newer houses in particular can get very warm on hot days, so a shady patch in the garden can be a welcome respite from the heat. Shade can also help you to grow a more diverse range of plants, allowing you to incorporate sun- and shade-loving plants.

Before creating shade in your garden, think about why you are doing it. Is it to ensure that you can sit outside on a warm day and if so will it just be you, or will your whole family want to join you? Are you creating shade so that your house doesn't get too warm? Do you want to shade a patch of your garden that gets too hot and where even sun-loving plants struggle to thrive? You also need to look at the aspect of your garden and make sure that you put your shade in the right place, bearing in mind that this will change slightly as the year progresses if you live in higher latitudes. Once you know why you want to create some shade, you can look at some of the following options.

Parasols

A simple parasol can do the job, particularly in a smaller garden, but, if you have ever experienced sitting under a parasol for more than half an hour or so, you will know that this involves regularly moving to keep up with where the shade lies. This is a good option, though, if you live in an area that rarely gets too much sun, or if you generally enjoy

Top, Shade can be a welcome relief in many gardens, making your space more enjoyable. *Bottom*, Easy, quick and fun, parasols can add shade that can easily be removed.

GROW A NEW GARDEN

warm weather. The benefit of a parasol is that you can take it inside in the winter and put it away on cooler days so you can enjoy the sun.

Trees

Trees naturally cast shade, and the variety of different trees available means that you can choose light shade, deep shade, summer shade or shade all year round. Bear in mind that planting a tree for shade will likely result in the tree (and therefore the shade) getting bigger year on year. Buying large trees can be expensive, so you might have to fork out or be patient. Choosing a fast-growing tree can help.

I love to plant trees near a south-facing window. In the summer, it can get really warm and heat up my house to near-intolerable levels. Then, in winter, a deciduous tree will lose its leaves, giving you the light you need.

Creating shade in the garden doesn't have to be elaborate, but you can certainly have fun with it.

Gazebos, arbours, canopies and arches

Just as they create privacy, these structures can also create shade. Ensure that you are putting them in the right place and consider the shade that they will cast elsewhere in the garden, not simply directly underneath.

Boundaries

You can develop the boundaries in your garden to create shade, such as planting a hedgerow along a chain link fence, extending the top of a fence with a trellis and climbing plants, or building a wall. Bear in mind that these structures will only ever cast shade on your garden

CREATING A NICE SPACE

Creating shelter from wind in your garden can help other more tender plants thrive.

for part of the day in the summer. When the sun is directly above, you will need to have shade that protects you from above.

SHELTERING FROM WIND

Developing your garden with the wind in mind can be slightly challenging, given that wind is such a destructive force. However, if you live in an area that is particularly exposed, taking some measures to mitigate against wind can dramatically improve the feel of your garden.

In much the same way that we used structures and plants to create privacy and shade, we can use the same to protect from the wind. However, we do need to bear in mind the survival chances of these structures. Things that are likely to catch the wind are generally not a good idea. Screens, gazebos and canopies can easily get broken and potentially cause havoc elsewhere in your garden if they are hit with strong winds.

Fences

Fences are often blown down in wind, so either opt for a particularly robust fence or one that won't catch the wind. If you are building a new fence and intend to plant a hedgerow next to it, consider going for a chain link fence or similar that will let the wind travel right through it whilst still ensuring that children and pets can't get out of your garden. If you do want a wooden fence in a windy area, consider posts made of concrete. These will make your fence sturdier and will allow you to easily change individual panels if needed.

GROW A NEW GARDEN

Walls

Walls are an excellent choice for reducing wind in your garden, providing they are in the right place. Bear in mind that some walls and fences can act to channel wind into your garden, which is rather unproductive. If you're going to invest in a wall, make sure to fully assess this potential before committing. Walls are an expensive but long-term option, so, if your budget is healthy and you don't want to worry about maintenance, this is a great option.

Hedgerows

A hedge can be surprisingly sturdy against the wind if you choose the right plants. Stout, thick-trunked hedging plants work well to reduce the effect of wind. To gain the full benefit and avoid destruction, these plants must already be established, so I recommend planting them outside of the windy season. You may want to provide them with some support whilst they are becoming established. If you do stake a hedgerow plant for the first year, take the stake away the following spring when wind is at a minimum. This will help the tree to detect the presence of wind and thicken its trunk accordingly, making it more robust.

Trees

Robust or well-established trees can break up a prevailing wind just enough to make it less noticeable in your garden. Planting a double row of trees can be particularly helpful as it filters more of the wind and offers the trees a little more protection. Bear in mind that trees can also be a hazard in the wind, particularly when they are standing alone or near the house. Trees are a great solution as long as you aren't expecting high winds.

Even well-established trees can fall in high winds, so exercise caution when using trees as windbreaks.

CREATING A NICE SPACE

To establish a tree in a breezy spot, plant it in mid-spring when the risk of high winds is reduced. Consider buying a pot-grown tree rather than a bare-root tree, as these will usually establish faster. You can support a tree with a sturdy stake, frame or guys. I recommend not relying on these too much, though, as they are not fail-safe, and they may inhibit the natural ability of a tree to grow sturdier. A tree will naturally thicken its trunk if it is allowed to rock in the breeze. So, a stake is a good idea for the first year whilst roots are growing but should then be removed once the tree is properly established. The other option is ground staking, which involves using wood at ground level to anchor the root ball. This can be done by securing planks parallel to the ground on either side of the trunk. Ground staking allows the trunk to thicken naturally and helps anchor the tree even before the roots are fully developed.

Ensure you prune trees when it is appropriate to do so (which will depend on the species), as this will make them more capable of coping with higher winds.

USING AWKWARD SPACES

Using awkward spaces in the garden can actually be quite fun. Bear with me on this one. They can present both a challenge and an opportunity. Most gardens have awkward spaces: the bit down the side of the house or a dry, shady corner under a large tree. You can try to find plants that suit these kinds of areas. Indeed, in some gardens you may struggle to find spots that aren't awkward. If you just have a few awkward spots in your garden, here are a few ideas of what to do with them:

Composting area

These aren't the prettiest of things, and you might not want to make a feature of them, but they would be perfect in an awkward spot, particularly if they are well-sheltered from heavy rainfall and wind.

Shed or other storage area

If you want a shed, you have to put it somewhere. And if you have an awkward area big enough for a shed, this could be a fabulous opportunity to have somewhere to store your gardening tools and support your newfound addiction.

GROW A NEW GARDEN

This awkward space behind the garage is a perfect spot for a little shed.

Containers
Containers or pots are a great way to fill an awkward area of the garden. You can use the spot to hide containers that are currently not looking their best or to grow plants that wouldn't enjoy the soil but wouldn't mind the other conditions.

Trailing, climbing or hanging plants
Depending on what kind of awkward spot you have in your garden, you may want to plant something in or near it just to cover it up. For example, I have a small space between one of my raised beds and the fence at the back of the house. I have planted a couple of climbing plants that should help to just soften that area and make it less noticeable.

Pond
Yes, you can absolutely build ponds in awkward areas of the garden. They don't have to be big or spectacular to help build your garden

CREATING A NICE SPACE

Left, This awkward space between two sunken trees in my garden was perfect for a pond, making the space more interesting. *Right,* The same pond a year later is looking more mature whilst we wait for climbing plants to green up the fence.

ecosystem. You can even create a container pond. If you're building a pond under trees, though, please bear in mind that you will likely need to cover it in some way in autumn to prevent it from filling with leaves.

INCORPORATING VEGETABLES AND FRUIT

There have been so many books written about fruit and vegetable gardening, and I don't want to go over all of that advice here. Chances are, if you want to vegetable garden, you will have already picked up one of those books. But if you're unsure of how you will incorporate a vegetable patch into your garden, here are some tips.

Most vegetables like to grow in full sun, with at least 6 to 8 hours of sunshine a day during the summer. So if you want to grow vegetables

CREATING A NICE SPACE

in your garden, make sure that you set aside a spot that will get plenty of light.

Vegetables are not tasty just for us humans; they are also particularly prone to pest damage. One of the best ways to mitigate this is by growing plants in the greenhouse before putting them outside. This way, you can put healthy, robust plants into the beds outside, and they should be able to defend themselves against most pests. If you have particularly voracious pests in your garden, such as rabbits, deer or other hungry mammals, your vegetables will need some protection even after they are established. The protection that you need to offer them will depend on the pest that you are facing.

A greenhouse can help you grow plants for your garden and is excellent for vegetable gardening.

Regardless, if you want to have a vegetable patch, a greenhouse is a good idea. A greenhouse will allow you to grow plants that are suited to warmer climates and you will be able to extend your growing season. The size of your greenhouse should be proportionate to the size of your vegetable patch. You don't need a large greenhouse for a small vegetable patch. So, consider adding a greenhouse to your vegetable garden. If you would like to grow flower seedlings for the rest of your garden, opt for a slightly larger greenhouse, as space is almost always the limiting factor when raising seedlings.

If you want to grow just a little bit of food, most fruits and vegetables can be grown in containers. This gives you the option to be flexible with your space. Since fruits and vegetables are generally quite thirsty plants, when they are grown in containers they will need regular

Fruit trees are an excellent, low-maintenance way of growing food in your garden.

GROW A NEW GARDEN

watering. If you have a season where you know you won't be able to give them the regular care they need, try growing some Mediterranean herbs instead, as they are very tolerant to periods of drought. Containers give you this flexibility.

One of the easiest ways to bring edibles into your garden is to plant fruit trees and bushes. If you are wanting to build some height into your garden and are looking at planting some trees, consider whether you could use a fruit tree instead of a purely ornamental one. Fruit trees will provide you with a dazzling display of blossom in spring, often the first bit of colour in your garden after winter. Later in the year, they will provide you with some fruit. Fruit trees are also an excellent resource for your local wildlife. You will end up having to share some of your fruit, but you will be bolstering your garden ecosystem and feeding hungry animals. Besides, most established fruit trees provide way more fruit than we could eat by ourselves.

Brassicas like kale can look wonderful in a flower border.

Herbs are also an excellent addition to any garden as they add fragrance, and certainly should not be overlooked as edibles. The feeling of being able to garnish a meal with herbs freshly picked from your own garden is simply sublime.

Don't forget that vegetables can also be grown in amongst your ornamental plants. I like to grow scarlet kale in my flower beds. The plant itself is highly ornamental, and I can harvest from them when I need. So you needn't do one or the other when it comes to growing vegetables. Courgette (zucchini) plants, with their giant foliage, look superb in tropical gardens, and herbs form the backbone of Mediterranean gardens.

CHAPTER 6

Lawn care

Lawns can be a little bit needy, but, since most of us like to keep at least a small bit of lawn, it's important that we know how to care for them. There are several challenges involved with lawn ownership; watering, mowing, adding nutrients, repairing patches and creating drainage are all regular lawn jobs. If you are getting your seasonal lawn care right, though, you can often circumnavigate the need for more work.

LAWN CARE THROUGH THE SEASONS

A little bit of maintenance can go a long way when it comes to a lawn, and not doing that maintenance can cause more work for you in the long run. As with most things in the garden, there are several different approaches we can take. Most lawn-care advice will recommend sticking to a very strict feeding routine. I've had a lawn for my entire life and never used a lawn fertiliser. If you do, the results are evident quite quickly, but bear in mind that this will need to be applied regularly. If you employ the natural systems in place to feed your plants as covered in Chapter 2, 'It all starts with soil' (page 31).

Spring
At some point in spring, the lawn will start to grow quite quickly. When you start noticing this growth, you can start to mow. If you have tufts of grass where the grass has grown unevenly over winter, set your mower to a higher cutting height. Be sure to mow before any reseeding or overseeding.

GROW A NEW GARDEN

Reseeding in spring can help fix patches in the grass.

Once you have mown your lawn for the first time in spring, you should gently rake it. Any bare patches revealed can be reseeded when the weather starts to warm up. I recommend adding a little bit of compost first and sprinkling some seeds on top. Press them gently into the soil by hand or by laying down a plank of wood and walking on it. Don't walk directly on the seeds, as you risk damaging them, but a plank will distribute your weight for a gentler effect. The compost will help with water retention and give the young plants some nutrients. You will need to keep watering the seeds regularly until they germinate as they need to be continuously moist. Be careful not to stand on grass seedlings, as they are delicate. If foot traffic is a problem, reseeding in autumn might be a better choice.

If your lawn is patchy all over, or the grass is quite thin, you can overseed. You will first need to topdress the soil with lawn-topdressing or compost. Add just a fine layer, not enough to bury the existing grass, and rake it into place. Then scatter grass seeds. Be sure to keep grass seeds watered regularly for the first few weeks to ensure successful germination.

If drainage is a problem in your garden, you can start to address this as soon as the ground begins to dry out in spring. We will discuss improving drainage further on in this chapter.

Summer

Mowing your lawn regularly through the summer will help it to thicken up. If you reseeded in spring, either avoid these patches or let them establish properly before mowing. Your lawn might need watering in summer, but take care not to water too much as the grass will become dependent on regular watering. It is far better to water deeply and less

LAWN CARE

regularly than a little and often because this scarcity encourages deep root growth.

Autumn

In September and October, you may want to consider overseeding or doing some patch repair. Grass will grow slower this time of year, but it will be easier to avoid walking on it when it is so wet and cold out. Rake up leaves that may fall onto your lawn, as they can choke the grass. This is best done when the leaves are dry, but wet leaves might be unavoidable this time of year. Unlike a leaf-blower, a rake will help to scarify the surface of the lawn and remove any thatch from the lawn as well. I add brown leaves to my compost or use them to make leaf mould, which is an excellent seed compost.

If drainage is a problem in your lawn, opening the soils a little can help water to drain and help reduce compaction. Open the soil by inserting a garden fork or broad fork straight down into the soil and gently pulling it towards you by a few inches. Not enough to break the soil, just to get some air into it.

Winter

Lawns need little care during winter when the grass is not growing. Make sure not to mow it during this period and stay on top of any fallen leaves, removing as many as you can. It is easy for compacted soils to get water-logged during winter, so if this is happening, you may want to

Raking leaves from your lawn in autumn will prevent the grass becoming smothered, and the leaves can be used to make leaf mould for seedlings.

aerate your soils gently using a garden fork to allow for some drainage. If it is a recurring problem, you might want to consider adding some permanent drainage solutions to your garden (see the 'Flooding and drainage' section on page 146).

COMMON PROBLEMS WITH LAWNS

There are several problems that lawn owners can face. Lawns are generally not easy to care for, and getting an immaculate lawn can be a difficult and high-maintenance endeavour. Lawns are quite far removed from natural environments. Here are a few problems that lawn owners may face, and how to approach them.

Patchy grass

The main problem with lawns is either that they go brown in summer or they become patchy. Going brown just means it needs watering, and most lawns recover fine. But there are several things that can cause patchy grass, including low nutrients, waterlogging, soil compaction and pests. Patchy grass can also be created by heavy foot traffic and dog urine. If you're new to gardening and you're not sure what's causing the patchy grass, you should start by doing a little bit of investigation.

Patchy grass is unsightly. Getting to the root cause can help.

Trampled. If the patchiness is in a linear formation, it could be due to foot traffic. If this is likely to continue, you may want to consider putting in a path or stepping stones. Just ensure that if you are adding a path or stones you aren't creating problems for yourself when it comes to mowing the lawn.

LAWN CARE

Dogs. If there are lots of yellow patches on the lawn or the grass has simply died off, this could be the result of dogs urinating on the lawn. There are plenty of solutions for this one, the simplest being to identify where the dog is 'going' and water it after your pooch has done its business.

Pests. If the lawn is very patchy and bare and the grass doesn't seem to grow thickly anywhere, this could indicate that a pest lives in the soil, such as cutworms (moth larvae) or leatherjackets (cranefly

Examining your lawn for cutworms and leatherjackets

Dig up a small portion of your lawn, around 10cm (4in) deep, and sift through the soil. If you come across a fat brownish-grey grub with sharp mouthparts on one end, your lawn patches could be a result of cutworms. I recommend treating your lawn with a nematode specifically for cutworms. If you find long, thin grubs or shiny red-brown pupae, you may have leatherjackets. You can get a nematode treatment specifically for leatherjackets, too. If you don't find any evidence of pests in your lawn upon closer inspection, you may have compaction, waterlogging or low nutrients.

Grubs, like this leatherjacket grub, can cause havoc on lawns.

GROW A NEW GARDEN

larvae). In which case, you will need to examine your lawn and take the appropriate action (see the 'Examining your lawn for cutworms and leatherjackets' sidebar on page 143). Or you may find that you have something eating your lawn from the top down, like deer. If shoots at the top are being nibbled by something, this isn't such a bad thing as it usually encourages more growth.

Dried out

Coming across areas that are dry and hard when you dig into the soil is often a sign that it is lacking in organic matter, which means water is

I added organic matter to my whole garden when we first moved in and then overseeded. The lawn is still a work in progress, but it is much improved.

LAWN CARE

not able to travel through all the layers of soil. If your lawn is regularly drying out because of intense heat, you may want to consider one of the following approaches to keeping it a little healthier:

Shade. You might not want shade on your lawn, which is entirely understandable, but it can be a long-lasting and eco-friendly solution. If this is something you would consider, reread the section 'Creating shade' on page 128. This adjustment can help your lawn suffer less from evaporation. But be careful not to plant trees that require a lot of water, because these will likely not establish well in the area and will likely exacerbate the problem by soaking up more water. Bear in mind that shade could also prevent rainwater reaching your lawn, so consider this when choosing what to do.

Irrigation. The obvious choice for tackling dried patches is to add some irrigation or manually water the lawn yourself. This can be effective but also quite labour-intensive or costly. I recommend collecting rainwater over the rest of the year using water butts (rain barrels) and storing it for summer if you are able. (Note that some states in the US don't allow rain barrels due to the spread of water-borne pests.) But this storage will reduce some of the ongoing costs. Try to water at dawn, before the sun becomes too warm and just before the plants need it most. Watering this early allows the water to drain into the soil before it evaporates.

Drought-resistant plants. If you are feeling brave, you might want to consider planting something other than grass. There are plenty of drought-resistant alternatives to grass. The species best for you will depend on your local environment, but it's worth doing some research if you're ready to make big changes.

Organic matter. Topdressing your lawn with a thin layer of compost can help to increase water infiltration and decrease surface evaporation. You will need to do this regularly – once or twice a year. Also ensure not to cover up the grass because this can cause it to die back.

Sit and wait. Most lawns that dry out over summer will recover almost immediately when it starts to rain. So, if you're able to be patient, you can just wait for it to spring back to life when the rain does finally come.

FLOODING AND DRAINAGE

Waterlogged lawns aren't fun. So, let's take a look at how we can better create drainage in your garden to avoid your lawn becoming a soggy mess in winter and killing off your grass.

Many lawns, particularly in new-build houses, suffer from drainage problems. This can either be a result of heavy clay soil or from compaction caused during the build process. Some lawns might not have a problem with drainage but are just in areas prone to flooding. In these cases, it's unlikely that drainage will make much of a difference, so be sure to assess what is causing your lawn to be wet before investing your time and effort into mitigating for it. To do this, you will need to try to find out where the water is coming from. It could be as simple as rerouting rainwater runoff from your house, shed or patio. You will also want to assess the soil, like we did in 'Assess your soil' on page 46. Waterlogging can seriously hamper a lawn's ability to establish properly and will continue to create problems unless drainage is properly addressed. Thankfully, there are a few things that we can do to improve drainage in our lawns.

It's important to first determine why your lawn is becoming waterlogged. The most common reasons are compaction, poorly aerated clay soil or having a garden that sits close to the water table (in other words, that is prone to flooding). The water table refers to the usual level of the ground water. If your lawn is well raised from the water table, it likely won't cause any problems. If your water table is high, or your lawn particularly low, it will probably flood quite frequently.

If you are starting from scratch with your lawn or your lawn is not worth salvaging, consider digging to add some drainage underneath the lawn. It is possible to improve drainage under an existing lawn, but adding a proper drainage system right away can forestall the need for ongoing care. Adding drainage underneath only works properly where the problem is the water table or simply too much water reaching the garden, and not compaction. Dealing with compacted soils or heavy clay soils will require a different approach.

Let's look at creating a drainage system to tackle particularly bad flooding. The type of drainage system that you install will depend on the size of your lawn and on the extent of the drainage problem. The aim is to try to strike a balance between improving the drainage and

A seepage box in the bottom of a large soakaway pit.

not creating an area that is too dry in the summer for your grass to flourish, or that needs continued and extensive watering. Here are some good options for drainage systems.

Soakaways

Soakaways can be square or in the form of trenches. They are quite easy to install, and you can build as many as you need to address the problem in your garden. However, they do need to be quite deep to be effective. In some areas where the topsoil has been removed, this can be challenging, and other drainage methods might be easier to install.

To build a soakaway, you will need to identify the lowest point in your garden. In some gardens a trench works well, particularly if it is along a boundary. In other gardens, a soakaway hole may suffice. Water will naturally pool in the lowest part of your garden, so this is where you should build your soakaway.

A trench is best if your lawn has a slight slope to it, or if there is a long portion that is most prone to waterlogging. Your trench should be at least 60cm (2ft) wide and 60cm deep. If you have grass already, cut the turf and put it to one side so that you can replace it once you're done. After digging your trench, fill the bottom third with large rocks

GROW A NEW GARDEN

or brick rubble; the next third with pea shingle, gravel or sharp sand; and then the rest with topsoil. The topsoil can be from what you have just removed to dig the ditch. To check that the replaced topsoil is level, place a wooden plank on the ground and walk over it. This will gently compact the soil and give you an idea of if the surface is level or not. If you have a lawn roller, this will also work.

A soakaway pit is built in the same way as a trench, but it should be around 1m (3.25ft) square in size to be effective. You can build as many as you need in your garden at low points and areas prone to waterlogging. These are then filled with stones as in the trench or with a plastic soakaway crate.

French drains

A French drain is designed to prevent waterlogging and to carry surface water away from a building. It's simply a lined trench, backfilled with rubble or gravel, dug at a slight gradient away from the building, and usually includes a perforated drainage pipe. They are efficient and are ideal in gardens that suffer from waterlogging or even flooding.

You must start by digging a drainage channel that slopes slightly down towards a drain or soakaway. Line the channel with some filter fabric or woven weed matting so that there is additional fabric on both sides. Fill the bottom of the trench with gravel (about a quarter of the way), then add the perforated pipe and cover it with more gravel. Fold over the remaining fabric and cover with topsoil. You can then either replace the turf or reseed on top. Ensure that the fabric you use will allow water through so that you don't end up making the situation worse rather than better.

French drains are quite easy to install, but you should follow correct procedures for best results.

LAWN CARE

COMPACTED SOILS

Next, let's look at how to deal with compacted soils. Any soil can become compacted if it is subject to a lot of traffic, for example heavy machinery during the build of the house, or trampling from repeatedly walking on it. Clay is particularly prone to compaction. This is because clay particles are shaped like tiny plates. Unlike sand and silt, which don't completely fit together, clay particles can tessellate and form quite a dense, hard structure. This means that clay can get wet and slippery in the winter and hard and cracked in the summer. One of the problems with compacted soil is that it is lacking in organic matter. With the addition of organic matter, the organisms in the soil can help to decompact it and build better soil structure. Soil structure that is built in this way is quite resilient and won't easily become compacted even if you continue to walk on it regularly. But the process does, of course, take some time. Let's have a look at some of the ways that we can deal with compacted or clay soil.

Manual aeration

To manually aerate the soil, you will need either a garden fork or a coring aerator. The latter simply takes cores (that is, spikes or plugs) out of the soil, meaning that there is less topsoil under the lawn. You can use the spikes/plugs to fix any sunken areas of your lawn, or you can add them to your compost pile.

If the soil was compacted due to a completed event, like building works, this process can get air down into the soil to help the microbes recover and start building up the soil themselves. However, if your soil is particularly waterlogged, the problem with relying on aerating with a corer or fork is that it generally needs doing once or twice a year because there won't be much microbial activity in the soil. After all, even the life in the soil needs air. So I urge you to be gentle and try to do this only once, as part of a combined approach to aerating your lawn, particularly if using a corer and removing some of the topsoil.

Sand

One of the traditional approaches to lawn aeration is to add sharp sand or horticultural gravel. In this approach, we would first aerate the lawn and then sprinkle or rake in some shop sand or gravel. But again, this is something that is usually done every year in problematic areas. The trouble with this approach is that clay has an uncanny ability to fill in the cracks. So even if you are creating aeration with sand or gravel, clay will likely find its way into those spaces and leave you with the same problem again next year. If your aim is to change the entire composition of your soil by adding enough sand to stop clay from being the main component of your soil, this is almost certainly folly. Unless your lawn area is very small, this is unlikely to work and might end up costing you a fortune.

Gypsum

Gypsum is only useful as a soil amendment for fixing slightly salty soil. See 'Gypsum' in the section 'Soil amendments' on page 57.

Topsoil

The good thing about adding topsoil is that it can raise the level of the lawn and create a top layer that is well aerated, provided you are not planning to walk on it too much. If your lawn is mainly for show, this approach can work quite well. However, if your lawn is likely to experience a fair amount of foot traffic, the topsoil will also compact over time.

The benefit of this approach, however, is that it gives grass time to establish. If your lawn has a good coverage of grass, there will already be some organic matter added into the ground, little by little, on a constant basis as the grass grows and dies. The grass roots will draw up water, and the entire area will be much more manageable. Given time.

If your lawn is particularly large or all the grass is already established, adding topsoil might not be the best approach. Certainly, you can spread a thin layer where you want to seed the lawn, but avoid covering existing grass with topsoil.

Compost

Compost can be a good addition to a lawn as it will help aerate the soil, build natural soil structure and feed the organisms in the soil. The benefit is that the structure built in the soil will be resilient and cope well with lots of footfall. However, this approach can take some time because it relies on the processes in nature to help. See Chapter 2, 'It all starts with soil' (page 31), for a full breakdown of how this works. The other drawback with this option is that the surface of the soil, where

LAWN CARE

there isn't grass, becomes quite black, so if you have pets, or children who don't wipe their feet well when they come back into the house, it can create quite a mess.

You can spread compost over existing grass, and then overseed the lawn with fresh seed. This works quite well but you may need to repeat the process several times before the soil is fully decompacted.

Strong, deep-rooted plants

You can improve drainage by using plants. But grass isn't ideal for creating drainage because it tends to be quite weak and, once it becomes waterlogged, it dies off quite quickly. That being said, if you have well-established grass with a strong root system it will automatically improve drainage in your garden. Grass will suck up some of the water, decompact the soil with its roots, protect it from further compaction and help to feed the life in it.

Sunflowers are deep-rooted, easy-to-grow annuals that can help add organic matter to clay and increase drainage.

But there are plants that are much better suited to sorting the soil out for you. If you have a little time, you may want to consider this approach, especially if the area in question is not needed for use for a little while, or if you have tried a few solutions already without success. Plants such as alfalfa and sunflowers will send down deep roots that will help to decompact the soil and will add organic matter for just the cost of some seeds. I recommend sowing these seeds in late spring and giving them a season to do their thing. At the end of the season, you can cut the plants down to ground level and allow the roots to rot down into the ground. Whilst this approach may not solve all types of drainage issue, it's easy, fun and cheap, so you don't have much to lose.

◆ ◆ ◆ ◆ ◆

Fixing drainage issues in your garden can be an ongoing challenge. With increased rainfall forecast as a result of global warming, the challenge may get worse. It's certainly worth putting in some regular effort if you want to keep your grass looking good.

As with most things, the best approach will always be one that is well considered and combines several different techniques. You might be in a rush to fix the problem but, if you don't know what is causing it in the first place, you could be wasting valuable time and money without getting any results.

ALTERNATIVES TO A LAWN

If I'm being completely honest, I'm not a huge fan of lawns. They are relics from history. Lawns first came about in the eighteenth century as a means to show off wealth. In those days, land meant food. It was a grand statement to have large swathes of land that weren't being cultivated to produce food. A preserve of the wealthy. The bigger your lawn, the more money you had. Lawns were considered a waste of space. Nowadays, it is ingrained in our psyche to want a well-tended and neat-looking lawn. Some of us never even question why we want that. For me, I do keep a lawn but as environmentally as I can: limiting irrigation, keeping slightly longer grass and introducing other species, such as daisies and clover. There are plenty of alternatives to lawns, including patios, decks, gravel, sand, and just planting the whole lot.

Of course, there is still a lot to be said for lawns. They provide us with somewhere to sit, lie or play in our gardens. We use them for summertime barbecues or paddling pool fun for the kids. They give our pets somewhere to run around, and our kids somewhere to play football. They do play an important role in many of our lives. But as we have just seen, lawns can require quite a lot of maintenance. In fact, many of us will find ourselves doing more maintenance on our lawns

LAWN CARE

than on the rest of our garden put together. Luckily though, there are some alternatives to grass, if you're feeling adventurous.

Moss

Bear with me on this one, as I know that many people fight hard to rid their lawns of moss. Moss exists in damp, low-nutrient soils, so if you don't significantly improve the state of your soils (by adding organic matter) you will continue to get moss. The thing is that moss is actually hard-wearing, doesn't need to be mown, feels lovely underfoot and can create a very even-textured, green surface when left alone to do its thing.

It might not be particularly traditional in most western countries, but in countries such as Japan they painstakingly remove the grass so as to not upset the moss. So, if you have a lot of moss and you're fighting a losing battle, take a moment to consider why you don't like the moss in the first place, and perhaps find your peace with it. You can buy Irish moss, which spreads well to fill in gaps where grass struggles to grow.

Moss is a wonderful alternative to grass.

GROW A NEW GARDEN

Chamomile lawn

Lawn chamomile (*Chamaemelum nobile* 'Treneague') is an adorable, low-growing and non-flowering plant that is a great alternative to grass. It requires a well-drained soil and plenty of sun and works best in areas without a lot of foot traffic. Lawn chamomile is excellent for wildlife and its soft feathery leaves are lovely underfoot.

Euromic micro clover

This clover is specifically bred to be low growing with small foliage and very few flowers. It can create a nice, even, green coverage in your garden, and helps to improve nitrogen levels in your soil. It's easy to grow from seed and will tolerate a wide range of conditions.

Meadow

If you want to attract wildlife to your garden, growing a meadow instead of standard grass can be an excellent choice. Which meadow mix you use will depend on your geographical location and what type of plants are likely to grow well in your garden. Bear in mind that the meadow is not hardwearing against foot traffic and can play host to some insects that you don't necessarily want near your home, such as ticks and stinging insects. So, it's not a great choice for families, but you could certainly use it for a portion of your lawn if not the whole thing.

A meadow is a beautiful alternative to a lawn but not practical if you want to regularly walk on it.

CHAPTER 7

Basics of plant care

I am a firm believer that having a good understanding of something can forestall many questions and problems. For example, a rudimentary understanding of nutrition (combined with a willingness to heed that advice) could prevent many unwanted medical conditions from cropping up. The same is true for gardening. If you understand what plants need, you can care for them more effectively and actively prevent problems from arising, rather than treating them once they are already a problem. Of course, we are not clairvoyant, so you will still need to treat some issues as they appear; but, if the basic needs are met, plants do a good job of fending for themselves.

SUNLIGHT AND WATER

Most of us learn from a young age that plants need sunlight and water. My children have presented me with beautifully cheery drawings depicting a large flower, a bright yellow sun and a rain cloud. Like many things taught to young children, it is of course an over-simplification. It's not untrue, but there are plants that can exist in deserts where there is almost no rainfall and on forest floors and caves where there is almost no light. These plants' ability to survive and thrive in these conditions comes down to adaptation. They have adapted to capture any sunlight they can or hold on to and use water in the most efficient ways. Nevertheless, sunlight and water are indeed basic plant needs. It's just the amounts that vary, and the same is true for many other requirements for life.

When you first add a new plant into your garden, you will need to firm the soil around it and water it in well. These two actions help the soil to fill in any gaps, which helps the roots get water whilst they are adjusting to their new surroundings.

OXYGEN

What most people don't learn at school is that all plants need oxygen for respiration. Yes, even aquatic plants. Oxygen is a basic requirement for almost all life on Earth. Thankfully, plants produce more oxygen than they use. Most plants that we encounter in our garden will take in oxygen through their roots. They use this for respiration in just the same way as you and I. We respire through our mouths and noses. If our faces were covered and we were unable to breathe, we wouldn't last long at all, mere minutes. But we can survive without water for a day, maybe even a week. The same is true for plants. They will die much faster when starved of oxygen when than starved of water. Of course, some plants have adapted for life in aquatic conditions and are able to extract dissolved oxygen from the water. Some plants can pull oxygen from water if their roots happen to be temporarily submerged, and different plants can cope with this submersion without drowning for differing lengths of time.

Water is a basic need for plants: when they are first planted, they will need more water to help them establish.

However, if you are trying to grow a standard garden or house plant, most of which are not adapted to semiaquatic conditions, you must ensure that your soil contains both water and air. Well-drained soil, like the stuff we discussed in Chapter 2, 'It all starts with soil' (page 31), will both hold on to water and contain lots of air. In fact, some soils with really good structure can contain more air than water, making them like a giant sponge.

BASICS OF PLANT CARE

NUTRIENTS

Some children learn that plants need soil to exist, which is not completely true. Some plants certainly do, but try telling that to the many people worldwide who grow plants hydroponically. Or to the *Buddleia* (butterfly bush) growing right out of someone's windowsill on my walk to my children's school. Or, indeed, try telling that to all epiphytes (plants that grow on trees and other plants, such as orchids and air plants).

Despite some plants not actually needing soil, it is still the most important part of your garden. I know I talk a lot about soil, but understanding it is key to successful fuss-free gardening. One thing that all plants have in common is nutrients, whether they get them from rainwater, a hose pipe or the soil. It's important to understand where nutrients come from naturally and how they are processed and absorbed by plants. So, if you didn't get that the first time round, go back and read the section on nutrients in Chapter 2, 'It all starts with soil' (page 31). In the soil there is a smorgasbord of micro- and macro-nutrients needed by the plant to support healthy growth, their natural immune system and their ability to adapt to different conditions in the garden. In most mainstream fertilisers, the three main nutrients are nitrogen, phosphorus and potassium (N-P-K). This limited set fails to recognise the crucial importance of micronutrients that contribute to

---

Q. I can't grow... (insert troublesome plant name here).
A. If there is a plant that you really love but you can't grow it in your garden, chances are the conditions aren't right. You can try changing the conditions (this is easiest to do within a pot or container), or you can try a different variety of the same plant. Sometimes when plants are selectively bred for colour, shape, scent or other attributes, they can lose some of their toughness. If you can find out what the original species of the plant is and find a variety more closely related to that (it'll be one of the oldest cultivars), this variety will probably be more robust and might be a viable alternative.

their flavour as food, self-defence, colour and overall plant health. These others are called *micro*nutrients because they are needed in smaller quantities even though they are no less important. Feeding your soil with organic matter helps to ensure that your plants get a balanced diet, including all the micro- and macronutrients.

GROWING PLANTS FROM SEED

There is a tongue-in-cheek saying in the gardening world that goes, 'Friends don't let friends buy annuals.' This is because annual plants are one of the easiest groups to grow from seed and buying them as started plants is a bit of a waste of money. There are, of course, plenty of decent reasons to buy annuals: all your seedlings were eaten, you forgot to sow them and now it's too late, you didn't have the space to sow them but do later in the season, and so on. Annuals include many flowering plants and vegetables. Despite the ease of growing them from seed, many people buy them as young plants. I completely understand this if you are busy or lacking space and just want an easy pop of colour. But, if you have some space and you want to keep the costs down, one of the best ways to do so is to try to grow things from seed, starting with annuals. Many seeds you can sow directly into the ground where you want the plants to grow – the hard part is remembering you put them there and not weeding them out by accident!

Sowing seeds is so easy, but many people shy away from this job on the grounds that it seems tricky. But say you have one plant that you particularly like to grow in your garden and you have to buy it every year – why not try growing it from seed? What have you got to lose? If it doesn't work, you can simply return to your original plan and go out to buy some. If it works, however, you will have all the plants you need for a fraction of the price, and you can collect the seeds at the end of the season and sow them again next year.

Equipment for sowing seeds

Sowing seeds is easy and simple, and you can usually do it using household items. Here's what you will need:

Seed trays. You can sow seeds into trays, pots, old toilet roll inserts, module trays or straight into the ground. Basically, if it holds

BASICS OF PLANT CARE

compost and allows water to drain, it can be used for seeds. If you are having problems with seeds rotting, chances are you are watering the compost too much, there is not enough drainage in the container, or the seeds have been sown at the wrong time.

Seed compost. This material is generally a well-drained and fine compost with not too many nutrients. Your seedlings don't need too many nutrients, and a rich compost may produce uneven or unhealthy growth. So, if you can, always opt for a seed compost. If you want to make your own seed compost, leaf mould is the best choice. Before dropping their leaves, trees will draw back the nutrients from the leaves. This habit means that autumn leaves are generally low in nutrients, so composting them makes an excellent seed-sowing medium.

Seed compost should be fine and not too rich; and it can be sieved if it is lumpy.

Compost sieve. If your compost is lumpy, small seedlings may struggle to push up through it. You can resolve this by using a sieve. You can make one yourself using a wooden frame and some fine metal mesh with holes of around 4 to 5mm.

Fine watering rose. You don't want your seeds to float away or become too wet when you water them, so it's a good idea to use something that will spread the water out. I recommend using a watering can with a fine watering rose. If you don't have one, you can take a clean plastic bottle filled with water and pierce some holes in the lid using a drawing pin. When you turn it upside down and give it a gentle squeeze, the fine application is perfect for seeds. I opt for this over a watering can when I'm working with very small seeds.

GROW A NEW GARDEN

How to make leaf-mould seed compost

Collect fallen leaves in autumn. These can be from your garden, other gardens or public spaces, if they are likely to be collected up anyway and it is legal. Note that in public spaces you should collect tree leaves rather than leaves from bushes and shrubs because tree leaves are less likely to contain dangerous contaminants. They are further from the ground where any fertilizers or pesticides might be sprayed and from the direct output of car exhaust. Put the leaves into a large plastic bag or reusable bin. Store them somewhere that will keep them moist but will not allow them to become waterlogged in rainfall. After one to two years, you will have a fine brown compost that is ideal for seed sowing.

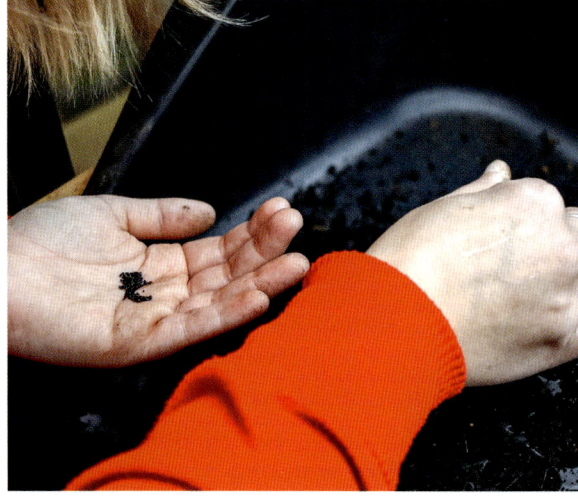

Sowing seeds is easy and fun, and a much cheaper way to grow plants.

Seeds. These can easily be collected from plants and saved, or you can buy or swap seeds. But they do need to be stored correctly. Placing a sealed box (so that pests can't get in and eat them) somewhere cool and dry is perfect. Buying good-quality seeds can be a little bit hit-and-miss. Generally, you get what you pay for when it comes to seeds, so if you want a better-quality seed, go with a well-known supplier and don't go for the cheapest options.

Warmth. Some seeds, such as chillies, need some warmth to germinate. If it is too cold, they could end up sitting in damp compost doing nothing and, eventually, they will rot. As a rule, it's a good idea to sow seeds somewhere that is warm but not hot. Warmth

BASICS OF PLANT CARE

Testing seed viability

If you are unsure if a pack of seeds is viable, you can perform a simple test to check. Start by thoroughly wetting a piece of paper towel and then laying it in the bottom of a sealable plastic box. Carefully lay a few of the seeds on top of the kitchen towel and close the lid. Place the box somewhere warm, where you won't forget to check on it. After a few days (depending on the germination time of the seed in question), inspect the seeds. Viable seeds will have started to crack open and produce tiny white roots. These first roots are called the radicle. If your seeds have roots, you can gently lift them from the kitchen towel or cut a square out of the kitchen towel and plant the whole thing in seed compost. However, if after two weeks in the box the towel is still damp and there is no sign of life (not even cracking), the seeds may have gone past their best.

will speed up their growth, but if it's too hot it will cause too much evaporation from the surface of the compost, drying it out too quickly. Good ways to give your seedlings warmth include sowing them on a sunny windowsill or in a greenhouse. An electric propagator will warm up the compost and generally comes with a lid to stop water loss through evaporation. A heat mat can also work, but be careful to use one that can get wet because not only can seed trays leak but accidents can happen when watering.

A make-shift greenhouse

Clear plastic storage boxes with lids also work well as make-shift greenhouses. All you need to do is find somewhere sunny where they won't get knocked over or disturbed. Place the lid down first, if it is opaque, and put your seed pots or trays onto the lid. Then turn the clear plastic box upside down and clip it into the lid. This setup will keep your seedlings protected from pests whilst also ensuring that they

If you enjoy growing plants from seed, consider investing in a greenhouse.

have warmth and light to grow. Make sure to check on your seedlings regularly, and let some air circulate through on warmer days by propping open the box with a stick or opening the lid.

TIPS FOR SUCCESSFUL SOWING

The only thing that most seeds need for successful germination is the presence of water. We have all tried that experiment where you sprout seeds on a wet cotton ball or a paper towel. Water breaks down the seed coat, which releases the embryonic root (the radicle). The radicle then sources water, which helps to put up the first seed leaves or leaf

(the cotyledons). Some seeds are sensitive to heat, and some require light to germinate. It is always best to follow the instructions on the seed pack for best results.

Here are some top tips for successful seed sowing:

Choose a fine seed compost or use a sieve. Making your own mix of composts can work well, but make sure there aren't any big lumps, which can get in the way of roots and shoots.

Depending on the seed, you should usually avoid extreme temperatures, which could hamper germination, affect water availability and even damage emerging seedlings.

GROW A NEW GARDEN

Always plant at the suggested depth. Some seeds don't contain the energy needed to push through deep compost. Some need light to germinate and will only do so if sown on the surface.

Larger seeds need larger pots. This may seem obvious, but some seeds produce very large seedlings compared with others. Choose a pot size that is proportionate to the size of the seedling. If the compost is drying out too fast, move up a pot size.

Know when your last frost date is and sow your seeds accordingly. This will ensure that you don't have to look after huge plants inside your home or greenhouse and that you don't put tender plants out before it is safe to do so. Your last average frost date might surprise you! Late frosts can be devastating if you're not expecting them.

Successful germination doesn't need to be complicated.

A note on peat-free compost

Peat is an incredibly valuable resource. It helps reduce flooding in upland areas, it is home to some fabulous and rare biodiversity, and, despite only covering 3 per cent of the Earth's surface, it holds more carbon than all the world's trees. It definitely does not belong in our gardens.

But peat-free compost can come with some challenges. Gardeners who are used to using peat will be familiar with its soft texture, its uncanny ability to hold water and yet be free draining, and its light consistency. Peat-free compost, however, differs from brand to brand.

BASICS OF PLANT CARE

Some are high-quality and others not so much. When choosing a peat-free compost, go with a reputable company, look for recommendations from other gardeners and check the colour. Dark brown, almost black, is usually a good sign. A fine texture is also desirable but difficult to achieve in bulk, as compost companies aim to do. The difficulty in getting a fine texture is because the compost has to be screened to be fine. Screening involves running it through a giant sieve. If the mesh of the sieve is too fine, it clogs up quickly and is slow and costly. If the mesh is too large, it can let through particles that are too large. This limitation is why I always recommend sieving your compost before using it with small seeds.

One potential pitfall with peat-free composts is the presence of aminopyralid (as we discussed in the section 'Farmyard manure', page 62). The key signs of aminopyralid poisoning are a twisting and curling of the leaves and, in some plants, a discolouration. However, these things can also be caused by extreme temperatures. Aminopyralid can last in the soil for up to seven years, so it can wreak havoc in gardens. If you are using a compost that contains animal manure, there is always a risk that it could contain aminopyralid. If you are unsure, there is a simple test that you can perform to check the compost you are using. Fill a pot and plant some broad beans in it. If the beans come out perfectly normal, you're OK. If they look like they have been cursed by the evil sea witch in *The Little Mermaid*, take the compost back and demand a refund.

There is, of course, one way that you can guarantee nice-quality compost that definitely doesn't have any manure or nasty contaminants: making your own! Making compost is so rewarding and fun. It helps you make use of the waste from your garden and kitchen and saves you money.

MAKING YOUR OWN COMPOST

Making compost is pure joy. Trust me, if you have no interest in making compost, by the end of one year of giving it a go you will be obsessed. I've had people say to me, 'Never in a million years did I think I would get excited about compost, but now here I am.' It's one of those things that doesn't seem very sexy, but opening your compost bin to find beautiful home-made compost teeming with worms is magical. There has been a lot written about composting. It can be quite complicated, and some people insist on trying to complicate it even more. If you are a compost manufacturer or a professional grower, complexity might make sense. However, if you are an amateur gardener, composting can be so simple that I promise anyone can do it. If you have tried composting before without any success, allow me to impart a little information that might help you along your way. If you have never tried composting before, this information will help you get started.

Before you get too enthusiastic and run outside to set up a compost system, I would just like to mention vermin. Compost areas can attract vermin from time to time, particularly if you are disposing of a lot of food scraps. It is recommended that you put your compost area as far from your house as possible for this reason. It is not possible to completely avoid vermin, but you have a better chance if you use a more enclosed system, such as a compost bin, instead of a simple compost pile. A hot composter, when at high temperature, will certainly not attract vermin. Alternatively, you can opt to use your kitchen scraps in a different way, for example using a *bokashi* composter.

Compost must be a mix of green and brown matter. Green matter (essentially that which still holds its colour: food scraps, grass clippings, green leaves) is rich in nitrogen and feeds the bacteria in the compost. The addition of green matter will help

At certain times of the year you will have more green matter than brown, so you must continually try to balance it out.

BASICS OF PLANT CARE

to generate heat in your compost, which is a sign of the necessary microbial activity. Breakdown of organic matter will be fastest when temperatures in the compost are between 35 and 60°C (95 and 140°F). Microbial activity generates heat but microbes will also become more active in increased heat. Increased temperatures also help to burn off weed seeds and kill pests that are unable to withstand high temperatures.

Brown matter (as you might guess, this is typically brown things: woodchips, cardboard, shredded paper, sawdust) helps stabilise the compost and gives the fungal decomposers something to feed on. In the absence of enough brown matter, the compost will become sludgy and wet. When a compost gets too sludgy or wet, it will become anaerobic, meaning it lacks oxygen. Without oxygen, many of the microbes will die, and so there will be little breakdown of organic matter. So brown material is very important.

Before you get started you will need to decide how you wish to house your compost. Here are some of the options, along with their pros and cons.

Compost pile

The humble compost pile is, perhaps, the simplest way of making compost. In fact, for many people simply throwing their garden waste into a pile and then forgetting about it inadvertently makes them some compost. However, it is quite slow. The reason for this is that it doesn't usually hold in heat well, and therefore it can take several years for the organic matter to break down. What's more, because it will break down under quite low temperatures, it will often contain a lot of seeds, which could be problematic if you are using it in pots or as a mulch.

PROS
- easy to set up
- doesn't cost anything

CONS
- slow composting
- often contains a lot of weed seeds
- can look quite messy

GROW A NEW GARDEN

Compost bin

Compost bins are a good way of keeping your compost area tidy. There are even some reasonably attractive-looking wooden compost bins that work excellently. The difficulty with using a compost bin is that it can be difficult to turn the compost within, which means that composting can take slightly longer. If you are going to have a compost bin, I recommend having two. That way you can take compost from one and turn it over into the second so that was on the top is now on the bottom and vice versa. You can then leave this transferred material to compost down, adding any new material to the now empty first bin.

Compost bins are certainly neater than piles.

When removing compost from a compost bin, it is almost always easier to lift the entire bin off the top, turning it into a pile. Scooping compost out from the bottom of a bin can be challenging.

PROS
- tidy
- easy to get hold of

CONS
- difficult to turn to get air in
- difficult to get up to high enough temperatures for optimum decomposition
- can be hard to empty

Hot composter

Hot composters are fantastic for creating compost quickly, and for confidently getting rid of perennial weeds and weed seeds. They are

BASICS OF PLANT CARE

> **Q: Can I add weeds to my compost bin?**
> A. The short answer is, yes, you can; however, some weed seeds will survive composting, and some weeds will thrive in a compost heap. Plants that can regenerate from root fragments should be treated with particular care. If your compost temperature is high enough (between 35 and 60°C / 90 and 140°F), this should be enough to kill root fragments and burn off weed seeds. If you are unable to reach these kinds of temperatures within your compost bin, you can try a hot composter. Or you can start by drowning your weeds before adding them to a compost bin. To drown them, add them to a bucket of water and leave them for a week or so until they start to turn brown. They can then be safely added to your compost bin without the need to worry about the temperature.

essentially just insulated compost bins, so they act in the same way as a standard compost bin but they don't let the heat out, making optimum composting temperatures much easier to reach. They often contain a thermometer in the lid so that you can monitor temperatures as well. It is entirely possible to make your own hot composter. Whichever method you choose, it is a good idea to invest in a compost thermometer.

PROS
- can make compost quickly (as little as 90 days)
- if optimum temperatures are achieved, it can burn off weeds and their seeds
- easier to get a good balance of ingredients because they are generally more forgiving

CONS
- expensive, unless you make your own

Compost bays

If you anticipate a lot of compost coming from your garden, you may want to set up compost bays. These are a series of consecutive bays,

Compost bays are perfect for larger spaces.

usually a minimum of one metre cubed each in size, to hold compost in various stages of decomposition. They are usually handmade, though kits are available. Bays usually have a removable front panel that makes accessing the compost very easy.

    The idea is simple. Add your yard waste to the first bay. Once it is full, turn the contents over into the neighbouring bay. This swap moves the partially broken-down matter from the bottom up to the top, and what was on the top is now on the bottom. Flipping over the compost introduces air, which helps to bolster microbial activity and speed up the breakdown process. Now add any new material to the empty first bay. Once that first bay is full again, the compost in the second bay should be ready to use, so you can spread it in your garden and then flip the first bay to the second again. You can add more bays depending

BASICS OF PLANT CARE

on how much compost you are likely to generate (how quickly a single bay will fill), but two is usually enough for an average garden.

PROS
- cheap to set up
- high capacity
- easy to access and use

CONS
- not ideal for smaller gardens

GROWING PLANTS FROM CUTTINGS

Growing plants from cuttings is one of the quickest and easiest ways to fill your garden on a budget. You can either take cuttings from plants that you already have in your garden, or you can take cuttings from friends' plants. The most important thing is that you take your cuttings at the correct time of the year. If you're taking cuttings from a flowering plant, make sure that you do not take a piece of the plant that has a bud or a flower. The reason for avoiding plants in flower is that producing flowers is energy intensive for a plant and they can become somewhat single-minded in that goal. So, for the best chance of success, always choose a stem that will not flower immediately so the plant will concentrate its energy on creating roots instead of flowers.

The most important thing to know about taking cuttings is that, until a stem has taken root, it is susceptible to rotting. So, ensure

Taking cuttings is a great way to get the most out of your plant and fill your garden.

that the medium you are using is well-draining. You can also root cuttings in water. Generally, green-stemmed cuttings, such as verbena, do well with rooting in water, but woody stems, such as willow, prefer to root into potting soil.

Softwood versus hardwood

There are two main types of cutting that can be taken from any given plant: softwood and hardwood. Softwood cuttings are taken from the soft and flexible shoots of the new season's growth. Hardwood cuttings are taken from older woody growth. The types of cutting are only slightly different, but it is still important that you know which you are dealing with before taking the cutting.

For all cuttings, you will need:

- a sharp knife, snips or a pair of secateurs
- hormone rooting powder or gel (helpful but not always required)
- specifically, a square pot (to provide more support than a round pot)
- well-draining potting compost
- horticultural grit or perlite (to add drainage if needed)
- a lid to go over the pot, such as a clear plastic bag or half of a clear plastic bottle
- somewhere warm, away from direct sunlight

Softwood cuttings

Softwood cuttings are ideal for propagating perennials, climbers and softwood shrubs. Don't forget that there are other ways to propagate, but taking cuttings is simple and easy and won't harm the existing plant. Softwood cuttings are easy to take and root easily, so, if you're new to taking cuttings, start off with these.

There are two different types of softwood cutting: basal and nodal. These names sound a bit intimidating, but don't worry; the difference is simple. You can take cuttings from shoots sprouting from the bottom of plants, or you can take them from fresh growth at the top of the plant, depending on which is available. If you are taking a cutting from the bottom of the plant, it's a basal cutting; otherwise, it's a nodal cutting. To ensure you are taking cuttings from the current year's growth, softwood cuttings will need to be taken in spring or early

BASICS OF PLANT CARE

Most softwood cuttings are nodal, from the fresh growth, so we will start with these.

summer. Let's look at how to do these two different types of cutting now.

To make nodal cuttings

STEP ONE: TAKE YOUR CUTTINGS
Choose a couple of fresh green shoots from this year's growth that don't have any flowers or buds. For each cutting, take a length of between 5 and 10cm (2 and 4in). It's important to take just enough that the cutting has some energy to produce roots but not so much that it will lose too much water before setting roots. Moving down from the growing tip, make a cut just below a node (where a leaf or leaves attach to the stem). This junction is where there will be the most hormonal activity and, therefore, the best chance of producing roots. You can trim your cutting after you have taken it too if you need.

STEP TWO: PREPARE YOUR CUTTINGS
Remove the lower leaves so there are only leaves in the top half of the stem and the bottom half of the stem is clear. Dip the bottom of your cutting into hormone rooting powder or gel (some will need it, but some plants will grow fine without extra hormones).

STEP THREE: PLANT YOUR CUTTINGS
Fill your pot with well-draining potting compost. The idea is to create a medium that is damp enough to encourage rooting but not so wet that it causes the stem to rot. Add horticultural grit or perlite to improve drainage and aeration in your pot.

Make a hole in the compost in each corner of your pot and plant one cutting per corner. The corners of the pot will help to provide some gentle support to your developing cuttings before they get going.

GROW A NEW GARDEN

STEP FOUR: COVER YOUR POT
You need to make sure that your seedlings don't dry out and that a good level of humidity is maintained to assist with rooting. You can do this by covering your pot with a clear lid. Use either a clear plastic bag, a bottle with one end cut off or a propagator lid. Stand your pot somewhere warm but not too hot (generally cuttings prefer temperatures of between 18 and 25°C [64 and 77°F]), so inside your home is usually perfect. Keep your cuttings away from direct sunlight because this can cause them to lose too much water too quickly.

STEP FIVE: TEND YOUR CUTTINGS
Keep water levels consistent using a watering can with a fine rose or a spray bottle to mist water on the surface of the potting mix. Depending on the plant, you can usually expect to find roots two to six weeks later.

Alternative method for nodal rooting

There is another way to root nodal cuttings, which is arguably easier for beginners, and that is rooting in water. Take your cuttings and prepare them in the same way as above (steps one and two), but you don't need any rooting powder. Take a jar of water and stand your cuttings inside. Most plants will root within a few days using this method and have a good root system within a few weeks. You will need to change the water every few days to keep it fresh.

The pitfall with this method is that roots produced in water tend to be a little weaker than roots grown straight into soil. So, when you are ready to pot your cuttings, be careful not to damage the roots, and make sure to keep the compost damp whilst they are settling in, about the first four to six weeks.

To make basal cuttings

Choose a shoot that has some leaves unfurled already and doesn't have any flower buds forming. You will want to take a section that is a little longer than a nodal cutting, around 7 to 10cm (3–4in). It's easier to use a knife for basal cuttings, so take a sharp knife or scalpel and cut as far down as possible: take a small sliver of the woody base because this is where the roots will form. From here, you can treat your cutting the same as you would a nodal cutting.

BASICS OF PLANT CARE

To make greenwood cuttings

Greenwood cuttings come from plants that have a harder, wood base to their shoots, usually because the shoots develop slower. These greenwood cuttings are halfway between softwood and hardwood cuttings. You can treat them in the same ways as you would a nodal softwood cutting, but take a slightly longer section, usually around 7 to 15cm (3–6in).

To make hardwood cuttings

Hardwood cuttings are used to propagate most deciduous trees, climbers and shrubs. They are easy and usually quite reliable, though you should bear in mind that the plants you are propagating might be slightly slower growing. You should take hardwood cuttings whilst the plant is in its dormant phase, without any leaves, any time from mid-autumn to late winter, which is helpful because there is less going on in the garden.

STEP ONE: TAKE YOUR CUTTINGS
Select a strong-looking shoot that has grown in the last year (it should have a soft flexible tip and no flower buds). Remove and discard the soft tip by cutting just before a node (in the dormant season these will just look like kinks in the wood). Then take a section 15 to 30cm (6–12in) long, cutting at a diagonal. This will serve the dual purpose of creating a larger surface area for water uptake and reminding you which end is the bottom. Cuttings will not work if you plant them upside down, and this is easily done when there are no leaves!

The only difference between basal cuttings and nodal cuttings is the way that you take the cutting itself.

Hardwood cuttings.

STEP TWO: PLANT YOUR CUTTINGS

Dip the diagonal end of your cuttings into some rooting hormone, and then you can plant them into pots or into a cutting bed. A cutting bed is just like a seed bed: an area with relatively fine soil and good drainage. From here, they will take root. When you start to see signs of growth on your cuttings, you can either leave them in situ or you can carefully dig them out and move them to their permanent locations. Take care not to damage the roots when you move them.

Whilst these general rules will work on most plants, some plants do have specific needs or methods when being propagated by cutting. So, I recommend looking up the technique for the individual plant if you are unsure. If no information is available, use the techniques listed here.

DIVIDING PLANTS

Another wonderful way to make more plants is to divide and conquer. This works excellently for perennials but can be a little bit nerve-wracking at first. Once you know how to do it and find your confidence, however, it is a great way to get more from your plants and fill your garden.

BASICS OF PLANT CARE

Dividing plants is done in one of two ways, either gently with your hands or using a large knife or sharp spade. The brutality of your weapon will depend on how tough the plant is to divide. The basic idea is to divide the root ball of the plant into several small root balls. Small root balls will have more difficulty overcoming potential pests than larger ones, so don't go too crazy and make them too small!

Some plants have nuances when it comes to dividing, all of which I can't cover in this book, but the basic process is as follows. If you are in doubt, look up the specific plant you want to divide and you should be able to find some tips.

When to divide perennials

Only divide perennials that you have had for a few years or bought as a large plant. I sometimes choose large plants from the garden centre because they can be divided into several smaller ones right away, which is much cheaper than buying several plants. Small plants will struggle to thrive as well as their larger counterparts, though, so only divide when they are getting to a good size.

As a rule, don't divide any plant whilst it is in flower; this will put too much strain on the plant whilst its energy is supposed to go into flowering. Additionally, don't divide your plant when weather or soil conditions are at their worst. Treat them as though you are introducing a new plant to the garden.

Most perennials are summer flowering and should be divided either in autumn once they have finished flowering, or in spring before they start flowering again. During the flowering period, plants concentrate all their energy on creating flowers and

A new hosta that is big enough to be divided already.

seeds, so root development goes on pause until this is done. Perennials that flower in spring, such as irises, can be divided in the summer after they have finished flowering. Perennials that flower over winter can be divided in autumn before flowering, or in spring after flowering is finished. For best results, I recommend dividing winter-flowering plants in spring.

To divide perennials

You will need:

- either your hands or a sharp spade, perennial spade or sharp knife (an old bread knife works well)
- a garden fork
- somewhere to put your newly divided plants
- some additional compost

Perennials should be divided just as you start to see the shoots come up in spring, or after they have flowered in autumn.

BASICS OF PLANT CARE

STEP ONE: REMOVE THE PLANT
Lift your plant gently using a garden fork. You should work around the plant, inserting the fork and gently pulling backwards. If it won't budge, move the fork back by a few inches so that you take up the entire root ball. Try to minimise damage to the roots.

STEP TWO: CLEAR THE ROOTS
You need to be able to see what you are doing, so shake off excess soil until you have a clear view of the roots.

STEP THREE: SEPARATE THE PLANT
If the root ball is loose and fibrous, as with hostas, you will be able to pull it apart. Make sure to choose an area between the stems to begin separating, and gently tease the roots apart. If your root ball is dense and hard, as in rhubarb, you will need to cut. To make a cut, choose a space between two stems and use your spade or knife to cut between the two sides.

STEP FOUR: REPLANT THE SEPARATE PLANTS
You can divide some plants into plenty of different plants, depending on their size, but remember that smaller plants are more vulnerable to pest attacks. By planting them together in clumps or keeping your plants quite large, you can mitigate against this to some extent. When replanting, add a handful of compost to the bottom of the hole, and give your plants some mulch on top, too. Make sure not to cover their stems or foliage.

STEP FIVE: WATER IN WELL
You will need to reestablish your plants, now, so give them some additional water over the coming weeks to help them recover.

CHOOSING HEALTHY PLANTS

When we first moved into our current home, there were two large garden beds between the patio and the lawn. Both were filled with soil and topped with slates. Apart from one large dock and a couple of box plants at one end, they were totally empty. Being something of a self-confessed plant addict, this hurt my heart. My need to be surrounded by plants and greenery borders on obsession. So empty beds simply wouldn't do. I set about sowing seeds in my new greenhouse

BASICS OF PLANT CARE

and taking cuttings where possible. There were, of course, some plants that I had to buy to get a bit of impact in the first year and to save myself waiting 10 years to see it grow. Buying plants is a lovely, and sometimes necessary, way to fill a garden quickly or to put some structure into your garden around which to plant smaller things you grow yourself. It is important, however, to choose healthy plants, so here are some tips. If you are plant shopping and find something you love but it doesn't pass the following tests, make a note of its name from the label and try to source it elsewhere.

A reputable source

When buying plants, it's almost always best to buy them in person rather than online. However, in some cases online is the only option, such as if you're looking for something slightly different or your local garden centre doesn't have a great selection. In such cases, you should do some research about the best, most local supplier and, if you can, order from a specialist nursery for the plants you want. Specialists will be able to offer you advice on how to look after your plant, and you can usually trust that they have looked after it well before it comes to live with you. If you're buying from a nursery or garden centre in person, take a look around. Are there dead or dying plants? Do you see yellowing leaves, or wilting? These are signs that the plants aren't being well cared for and should serve as a warning to go elsewhere, or to take extra care when choosing your plant.

Leaves

Inspect the leaves of the plant. If there are any yellowing leaves, this is a sign of a potential problem that the plant might not be able to recover from, such as waterlogging. Brown tips to the leaves or physical damage can also be signs that the plant hasn't been well taken care of but may be signs of bigger problems that aren't yet visible.

Black spots on the leaf can be a sign of a fungal infection, such as rose black spot. You don't want this in your garden as it could contaminate other plants, so don't buy any plant with black spots. Plants that are wilting should also be avoided. Whilst they may recover, it is usually a sign that they haven't been well cared for.

If plants are already healthy when they arrive in your garden, they will have an easier time getting established.

GROW A NEW GARDEN

Potting mix

Check to see if the compost that the plant is growing in is particularly wet or dry. A layer of algae on the top is a sign that it has been very wet for quite a while. Don't forget, some plants enjoy lots of water, and others prefer very dry conditions; just make sure that the mix they are in matches their requirements.

Roots

Plants that have been sitting in their pots for too long will become rootbound. This can lead to a lack of nutrient availability, some roots rotting and the plant being generally constrained. In some cases, it can mean that the plant will never thrive, even when released from its pot. So, turn the pot upside down and check if there are roots coming out of the bottom of the pot as this is a sign of a plant being rootbound. If you are able, you can gently lift the plant out of its pot to check. A really rootbound plant will look as though it has woven its own basket from its roots, and these plants should be avoided.

    The other thing to look out for in the roots is the colour. Older

Top, This rootbound plant will struggle to thrive. *Bottom*, Healthy roots on a quality pot-grown tree.

roots are generally darker in colour. New roots are white with little hairs. Seeing white roots on a plant is a sign that it is in good health. If the roots are dark brown, this could be a sign that the plant has been overwatered and the roots are rotting away. Root rot could be terminal and should be treated as a red flag!

Bugs

An easy way to bring pests into your garden is with an infected plant. Pests like vine weevils, mealy bugs and scale bugs can easily go undetected in a new plant and end up spreading to your existing plants, causing a huge problem. So, before making any purchases, make sure to thoroughly check the plant for any signs of bugs and be prepared to walk away if you see any.

Flowers

When buying flowering plants, it's nice to know what the flowers will look like, and also what colour you can expect (plant labels aren't always accurate). But be aware that if you are buying a plant that is in flower, you will have missed out on some of its flowering time already. The flowers could finish within a few days, leaving you waiting until the following year to enjoy them. Where possible, buy flowering plants before they are in flower, or when they are in bud, so that you get the maximum enjoyment from their flowers in year one.

Weeds

Weeds are quite a common occurrence in plant pots nowadays, but they are a sign that the nursery hasn't been paying close attention to the plants. The weeds will also be competing with your prospective plant for nutrients and water inside the pot. (Let's not forget that potted plants don't benefit from life in the soil.) It is also entirely possible to bring weeds into your garden that you then can't get rid of, so it's always a good idea to check first.

KEEPING A JOURNAL

I cannot recommend keeping a journal highly enough. Some years you will have fabulous success with certain plants. The next year, the same plants could be complete failures. This predicament is normal in the

GROW A NEW GARDEN

Keeping a journal helps you remember what you have in your garden and records your successes and failures.

gardening world but, if you are keeping a journal, logging when you have planted and sown things, what the weather was up to, what the temperatures are like and any other details that might be important, you can start to figure out the key to your successes and failures. Over time, this will make you a better gardener. Also, it will help you keep track of what you have sown and planted. A gardening journal is a completely personal thing and I cannot tell you how to do it, but I can recommend that you give it a go.

CHAPTER 8

Managing pests and weeds

Pests can be extremely contentious. There are those of us who would rather pretend the pests don't exist and turn a blind eye to the damage, and those who wage a constant war. But some gardens are more prone to pests than others – why is that?

The key to good pest management in organic gardening is to create an environment where pests cannot thrive. This doesn't mean making it so hostile that nothing can survive but rather creating a balanced and sustainable ecosystem where pests are not the dominant life form.

IMPORTANCE OF A HEALTHY ECOSYSTEM

We have already talked about healthy ecosystems and how our gardens are ecosystems, whether we want to consider them one or not. By far the best way to manage pests in your garden is to create a healthy ecosystem that manages your pests for you. This option requires healthy soil and a variety of plants and habitats. The greater the variety in your garden, the greater the biodiversity. With increased biodiversity comes increased balance. There will be enough good species to help you manage and control the species that cause damage to your plants. The addition of mature trees, ponds, log piles, shrubs, bird boxes and feeders, bug hotels and sheltered, undisturbed spaces in your garden will all help to promote biodiversity and create a healthy and stable ecosystem.

The more you can do to encourage wildlife in your garden, the fewer pests you will find in the long term.

Our gardens are islands

In 2007 I was lucky enough to spend a year studying island biogeography in New Zealand. I had been inspired by one of my favourite books, *The Song of the Dodo* by David Quammen. The book is about the ecology of islands and how the fragmentation of habitats affects the species within them. It's a beautiful work of literature as well as a fascinating account of many examples of habitat fragmentation and island biogeography. If that sounds interesting to you, give it a read.

When I first started gardening, my study of island biogeography started coming back to me. The principles around newly formed islands and newly formed gardens are more or less the same. Most gardens, like smaller islands, are also unable to permanently house large animals. There simply isn't enough space or food for them. In some cases, this can mean an imbalance in the species that come below them on the food chain. If your garden forms part of a wider ecosystem and these animals are able to come and go as they please, this won't be a problem for you. But if you have a dog or some feral children, you're unlikely to leave gaps in your garden boundary big enough to let in wandering predators. You may also not want to let large predators into your garden, which I completely understand.

Please, bear in mind that I am writing this from the comfort of England, where our most ferocious predator is the fox. The point I'm trying to make is that sometimes it doesn't matter how diligently we try to create a balanced ecosystem in our gardens – they are not always fully capable of holding one. All this means for us is that we need to give our garden ecosystem a little help sometimes, some conservation work, if you will.

My own garden island, surrounded by houses on a new estate.

MANAGING PESTS AND WEEDS

CONTROLLING PESTS ORGANICALLY

Dealing with pests can be enormously frustrating. I have a healthy population of cutworms in both my garden and my allotment. They crawl through the soil and eat the roots of my young plants, including grass, resulting in huge losses of my carefully cultivated seedlings and in patchy grass. If I were remotely inclined towards destruction, a broad-brush 'kill them all and start again' approach could seem very tempting. There is a pesticide that you can apply to your lawn for the small cost of about £9 (around USD 12 at the time of writing). Whilst I have not looked into the other effects of this treatment, I would be prepared to

GROW A NEW GARDEN

A hungry slug squeezed in next to a raised bed to wait for dusk.

bet that cutworms and leatherjackets aren't its only victims within the soil. There are, thankfully, plenty of ways that we can reduce pests organically. I have combined my love of gardening with my knowledge of ecology to bring you my Three Golden Rules for organic pest control.

Rule 1: Know your enemy

Sun Tzu's famous advice in *The Art of War* (something gardening is often depicted as) is to 'Know thine enemy.' If your garden is suffering the effects of one particular pest, the first thing you should do is some research. Educate yourself about this creature's lifestyle and lifecycle. It might just turn out that you were inadvertently rolling out the red carpet for them.

HABITAT

What habitat does your pest enjoy? Do you have lots of that around your garden? Are you able to reduce the amount of it? For example, slugs like to hide somewhere they are less likely to be eaten in their sleep. So, they enjoy corners, walls and the underside of things, where they are protected on at least one side. Raised beds make excellent slug habitat, providing them with plenty of places to hide from hungry predators. So, if you do have raised beds, ensure the plants you are growing in them are large enough to withstand these pests, or plant something the slugs don't like to eat. You can also be vigilant and regularly check for and remove slugs, which can help prevent some damage.

LIFECYCLE

When is your pest active? Does it hibernate, or spend winter as eggs in the soil? It might be that you find one of these stages is easier to control than the others. For example, the cabbage white butterfly likes to lay its eggs on brassicas (like cabbage and kale). The butterflies and eggs aren't really pests, but the caterpillars are. You can help your brassicas by

MANAGING PESTS AND WEEDS

either covering them with some butterfly netting or being on egg watch during early summer when the butterflies are most likely to be laying. I brush butterfly eggs off the leaves when I find them; if they then hatch, the caterpillars are too small to climb back onto the plant and cause it damage. You should know that butterflies are more than happy to stick their ovipositors (egg-laying parts) through the net. This manoeuvre means you have to make sure the net is raised sufficiently from your brassicas to provide protection.

PREDATORS AND PARASITES

What eats your pest? Or kills it in other ways? Is this something you can research and also invite into your garden? For example, my nemesis pest – cutworms – aren't eaten by birds because they live underground (unless the cutworms find themselves being placed on the bird feeder by a frustrated gardener). They are sometimes eaten by moles but inviting those into your garden is kind of like the lady who swallowed a spider to catch the fly. They are, however, parasitised by nematodes, which (gross alert) crawl in through any orifices and infect the cutworm with bacteria, which kill the cutworm. The nematodes are species-specific and won't damage your soil ecosystem. If the grass is allowed to grow because of fewer cutworms eating the roots, there will also be fewer bare patches for the adult-stage moths to lay their eggs.

FOOD

What does your pest like to eat? Do they like young plants? Old leaves? And how do they access their food? For example, woodlice are a problem to some gardeners, but they aren't good at climbing and they don't have very strong jaws. Therefore, they tend to go for short, tender plants, especially if they have leaves dragging on the soil. An excellent way to prevent this kind of damage is to plant

Another predator is a parasitic wasp that lays eggs inside a caterpillar, which is eaten from the inside out. These are the pupae next to the hollowed-out caterpillar.

GROW A NEW GARDEN

things out slightly later in the year so they are a little larger and grow a little faster; then, be diligent about checking on your plants and making sure to remove leaves that are dragging on the soil.

Find out what pests you may face

If you don't know what pests you might face in your new garden but you'd like to be prepared, there are some useful things you can do to find out.

Slugs love to eat my favourite dahlias, leaving silvery slime in their wake.

TAKE A CLOSE LOOK

If you are starting with a garden that already has some plants in it, chances are that your garden ecosystem is better than if you were starting with a paved garden or a plain lawn. This is good, as it means you will likely encounter fewer problems. However, it is possible that your garden could have some of the trickier pests. For example, honey fungus (*Armillaria mellea*) will wreak havoc on your trees in an already-established garden but won't be present in a garden that has no trees.

When you are first observing your garden, it is prudent to carefully inspect what is already there for any signs of damage. If you can spot anything that looks untoward, it may give you a good idea of what you can expect to grow and what might not be so successful.

ASK THE NEIGHBOURS

If your garden is likely to be regularly visited by a mole, your neighbours will tell you about it! Similarly, if your neighbours see a lot of anything, they will be more than happy to share with you their tales and woes. If you have any neighbours with a proper garden, go ahead and ask.

FRONT-LINE PLANTS

When you are starting a new garden, you won't know what you're going to be up against in terms of pests. There is nothing worse than lovingly raising your favourite plants only to have them eaten overnight when you put them into your garden. Similarly, if you have invested money in these plants, you want to know that they're going to be able to survive what your garden has in store for them. So if you don't already know what you might be dealing with, front-line plants are a useful way to get to know the enemy, for example if you're starting with a bare patch of land.

My favourite front-line plant is lettuce. Because everything eats lettuce, the lens through which you see your pests will be slightly magnified. Lettuce is super easy to grow from seed and the loose-leaf varieties can be grown all year round. Violas are another great front-line option: they are also easy to grow from seed and you can usually buy them at your local garden centre. Violas add a pop of colour, too, whilst being super hardy against the cold. By planting out some seedlings that have cost you little in effort, time or money, you can better figure out what pests you might be up against. This approach isn't fool-proof, and pests change throughout the year. However, small, just-emerging perennials and new seedlings are the most vulnerable. So, if you can scope out the enemy with some throw-away plants, it might just help. The worst that can happen is you get some salad.

Rule 2: Treat the cause

The Second Golden Rule of organic pest control is to take the lesson you are being given. If your garden is suffering from a particular issue, what you are likely to be seeing is the symptom rather than the problem itself. Just like we can't see the cold virus but we can feel the headache, chills and sneezes. Quite often, the proliferation of a pest, weed or other problem is your garden's way of saying it needs help. For example, if a plant was wilting, you wouldn't simply prop the leaves up; you would listen to what it was telling you and give it some water. When you are doing research on your pest, you might come across some clues that will help you. For example, the cutworms that I have in my garden thrive on nutrient-poor, damp soil. Therefore, I am seeing cutworms because my soil isn't in great shape, and the drainage needs some assistance. These are things I am working on at the moment but that will take time. In the meantime, I can treat the cutworms with nematodes to make it less painful (like popping some paracetamol when you have a cold). In the long term, I hope to fix the problem by addressing the cause – by improving the soil.

GROW A NEW GARDEN

Quite often the cause is that your garden is not functioning properly as an ecosystem, in which case you will need time, again. However, as we have already learned, predators and parasites won't come to your garden if there is nothing for them to eat or parasitise, so sometimes you need to be a little bit clever.

BUILDING AN ECOSYSTEM WITH LESS SACRIFICE

Planting out your garden and watching it get eaten can be pretty soul-destroying. You won't be the first to have this happen, neither will you be the last. As we said before, ecosystems take time to build, so how can we build an ecosystem that will manage our pests in the long run without sacrificing all your beautiful plants in the short term? There are several ways in which we can do this, and, as with most things, a combined approach is almost always best.

SACRIFICIAL PLANTS

Sacrificial plants are those that you don't mind losing to the odd pest or plague. Like front-line plants but after the garden is already going, to take the brunt of any attack. They are often fast growing and should be more tempting as a treat to your pests than the plants you are actually trying to grow. In my vegetable patch, I grow nasturtiums. They aren't invasive in the UK, and they are delicious to our cabbage white caterpillars that wreak havoc on our brassicas. They are fast growing and you can chop back the damaged and unsightly material only to have them spring back even more vigorous within weeks.

Sacrificial plants such as nasturtiums can take some of the heat off the plants you want to keep.

The nasturtiums help to take the heat off my brassicas because the butterflies will often opt to lay eggs on the nasturtiums instead of the kale. The best thing about using sacrificial plants, as opposed to chemicals, is that you are building the ecosystem in your garden. Sacrificial plants will help you to grow what you want whilst still making the prey species available to any predators looking for a new hunting ground. You don't have to find specific species to complement the plants you are growing either. You can simply grow more of what you love, and protect – with either nets or cages – the ones that you want to keep. If the ones without protection do OK, you can probably do away with the protection.

Rule 3: Be patient

Here's the thing – all of this requires patience. You have got this far into the book, so that bodes well for you already, but don't forget that you can't expect to fix your problems overnight, even if you're using chemicals. In any new garden, when visited by a pest there is a cycle that we can expect to see. Ecologists call it 'boom-and-bust population dynamics', which is actually quite descriptive.

BOOM AND BUST POPULATIONS

Without knowing this nuance of how populations work in nature, your patience might wear a little thin. This information should give you some hope that your work is heading in the right direction. Here's what it looks like.

When a population of something first arrives in your garden, the chances are there will be a lot of food for them to eat (they will have arrived for a reason). And since they only just arrived, there won't be any predators or parasites yet. Their population will be under optimum conditions, and the chances are they will thrive. This is called a *boom*. After a certain period (depending on the specific conditions), food will become less available, and some predators and parasites will begin to show up. The pest population will start to go down. When the pest population starts to go down, there is less food for the predators, so they begin to move on, too. The plants start to recover, and the pest population can begin to regenerate. However, this time round, it won't take as long for the predators and parasites to come back; some might even have stuck around. The result of this is that even the best population doesn't reach the same crescendo it reached the first time round; rather it reaches a smaller peak and then starts to go down. This pattern continues, like a bouncy ball, and slowly the population bounce-backs get smaller. In time the population of this pest will be at a manageable level, where you probably won't notice it.

GROW A NEW GARDEN

Caterpillars in my vegetable patch

In case you didn't know by now, I had caterpillars on my vegetable patch. They are a real pest for us kale-loving gardeners. I love kale so much I grow it in my flower borders, too; it looks great and it's handy for a snack if you get hungry on the job. The last garden I tended was a bare patch of lawn when I first moved in. I set about creating a vegetable patch in this garden with the primary aim of feeding my young family during the COVID-19 pandemic. I was also given the distinct honour, amongst many others, of becoming a homeschool teacher, a job that, as far as I can tell, requires a reasonable amount of gin and a lot of patience. But that's another story. My then 7-year-old was at the time learning about food chains in her science classes. As an exercise, we had to think of some examples of food chains. This was easy for my daughter, who got all the inspiration she needed just by looking out the window. The vegetable patch I had started in March was thriving by June. There were carrots and parsnips, sweetcorn and French beans, plus neat rows of onions and a trellis covered in peas. However, when we hit the second week in July, disaster struck. A butterfly found my garden and laid a heroic number of eggs. (OK, it was probably multiple butterflies.) Either way, it wasn't long before there were caterpillars everywhere you looked. A vegetable massacre. My beautiful Tuscan kale was reduced to sticks. My cabbages looked like my grandmother's lace curtains. And on a still evening, you could actually hear the sound of millions of munching caterpillars. That year, we didn't eat a single brassica from our garden.

Some of the kale plants recovered admirably, and I let them go to flower the following spring. They billowed out clouds of simple yellow flowers that were attended by bees, wasps, hoverflies, moths and more bees. The early spring blooms were a rare treat for them when nectar was scarce. These kale plants provided a draw for one pollinator in particular: a tiny parasitic wasp called *Cotesia glomerata*. The white butterfly parasite.

The following year I diligently sowed and planted my brassicas. Without the money for nets or cages to protect them, I sent them off into the world. But that year the caterpillar plague never happened.

MANAGING PESTS AND WEEDS

There were a few caterpillars, yes, but there was also plenty of evidence of another force at work. The tiny parasitic wasp had wrought havoc on their population. The empty carcasses of caterpillars were strewn about my garden.

These wasps lay their eggs in live cabbage white caterpillars and, when the eggs hatch, the larvae eat the caterpillars from the inside out. When they have consumed their victim, they pupate into tiny yellow, silken cocoons that extrude from the caterpillar. The evidence that we see is a hollow caterpillar with a group of small cocoons next to it. This sounds like quite a grim way to go, but it is part of the circle of life. This cycle succeeds in killing a single caterpillar and producing a clutch of new caterpillar-killing wasps. These wasps will stay in my garden if there is plenty of food (nectar from brassica flowers) and places to lay their eggs; or not if there isn't. The best thing about it, though, is that there is a pheromone released by the cabbage white caterpillar during this consumption process that warns cabbage white butterflies not to lay their eggs nearby. The result is that my garden wasn't affected (much) by caterpillars that year.

My brassicas were entirely stripped in the first year by caterpillars.

I ended up leaving the house and the population study. But I'd be prepared to bet that the following summer, the caterpillar population would be healthy, but not to the extent it had first been. Over time, the numbers of wasps and caterpillars would have balanced, and the garden would have been a mostly munch-free zone.

There are plenty of brassicas you can use to attract pollinators and parasitic wasps to your garden. All you have to do is let them flower.

PITFALLS OF CHEMICALS

Chemicals can seem like such an easy way to rid yourself of a problem. Like a silver bullet, you can close your eyes, pull the trigger and the pest will be sprayed away. The trouble with this approach is that it doesn't take into account the problems that the same chemicals can cause in your garden. Even if they seem benign like the so-called organic chemicals, like neem oil. Even these chemicals create imbalances in our gardens, and knock-on effects elsewhere. The real problem is that without the pests, and particularly if we are poisoning our pests with the likes of slug pellets, for example, the predators will not be coming to your aid. They skip town because either the pests die out too quickly to provide food or the poison in the pests kill the predators who do eat them. This lack of predators means that you will likely continue having to use these chemicals. Over time, the cumulative effect of using chemicals will deplete the health of your garden. Whereas, if you take an organic approach, you may find that your first few years are a little painful, but after that you can expect the health of your garden to improve year on year, and for your workload to go down, giving you more time to enjoy your garden.

COMMON GARDEN PESTS

The pests that you experience most in your garden will depend on your location, your climate, the vegetation in your garden and, sometimes, your unique bad luck. This list is by no means exhaustive, but once you get the hang of controlling the most common pests you can do your own research into the more obscure ones.

Slugs and snails

What the damage looks like: All the leaves missing from a plant and just the stems remaining; large holes on leaves, usually from the outside in; silvery slime trails

Slugs and snails are a common problem in gardens, and one of the pests that I get asked about the most. Silver trails and large sections missing from your leaves are their calling card. Slugs and snails love untidy gardens as there are plenty of places for them to hide. They also like damp spaces, so the bottoms of walls and around raised beds

MANAGING PESTS AND WEEDS

Slugs and snails are pretty easy to spot in the garden, particularly at dusk.

where they can tuck into the grass or soil at the bottom is perfect. Having plant leaves trailing on the ground can often cause problems with slugs, so maintaining your plants well can immediately reduce the amount of damage caused. Choosing plants that are less palatable to slugs and snails can also help. Red foliage tends to be less tasty than green; and anything with soft or tender leaves will be easier for them to eat. These plants may need a little extra protection.

If you find that your seedlings are being demolished by slugs or snails, you may want to start your plants in a greenhouse or on a sunny windowsill until they are a little larger. When temperatures start to rise in spring, and there are more hours of sunshine, plants will grow faster. What this growth spurt often means is that they are more able to defend themselves from pests such as slugs and outgrow the damage.

Regularly checking for slugs and snails around your pots and trays can also help to reduce damage. You can collect up your slugs and snails and relocate them. Be aware that slugs can travel back to your garden (particularly if they have their eyes on your hostas!), so I recommend taking them at least 20m (65ft) away and depositing them somewhere that won't make you any enemies!

Aphids

What the damage looks like: Flowers developing poorly; usually before you notice any damage, the actual aphids on the plants in green, black, brown or grey, most commonly at the top of plants near flower buds.

Aphids are sap-sucking insects that generally appear in spring and concentrate themselves around the soft new tissue of a plant, like a plant's buds. But they are easy to manage organically. Remember to check your plant and either wipe them off by hand or use a spray bottle to knock them off your plant with water. Once on the ground they

Left, These aphids are enjoying sucking the sap from a rose bud. *Right,* Some mammals, like these deer, can cause havoc in gardens.

won't be able to get back up onto your plant, and they will be easy pickings for predators.

Caterpillars

What the damage looks like: Large sections of the leaves missing, either from the inside of the leaf, working outwards or from the outside in; blackish-green poop underneath the damage.

Caterpillars can appear in vast numbers in the garden, but they are tasty to many creatures. They are also regularly parasitised by wasps. If your plants are healthy and growing in healthy soil with plenty of access to a full suite of nutrients, some will be able to create chemical compounds to attract parasitic wasps. Attract birds into your garden by having bird feeders and bird houses to help control caterpillar populations.

Nets can create a physical barrier to prevent caterpillar damage. By netting a few plants and leaving others, you can create a balanced ecosystem over time, without compromising the plants that you want to protect. Bear in mind if you are using nets: they can trap birds from time to time, so it's important to check them regularly and ensure that they are well secured. Butterflies also have no problem laying their eggs through nets if they can reach the plant, so make sure the net is raised above the level of the plant.

Mammals

What the damage looks like: Varies; small animals – eaten seeds, preventing germination; other animals – eaten leaves, bark or shoots.

MANAGING PESTS AND WEEDS

Mice, rats, rabbits, deer and other mammals can be voracious feeders in your garden. These kinds of creatures will require physical barriers to prevent damage. In the case of larger mammals, fences and guards can be used. Smaller mammals can be troublesome because they can burrow underground. Growing particularly susceptible plants in containers can help, as can using shelves to create a little distance between the pest and your plants. Strong-smelling plants such as chives next to your precious plants like lettuce, can also help to deter small mammals.

Cutworms

What the damage looks like: Plants wilting (they have been severed from their roots); patchy lawn.

Cutworms are a type of moth caterpillar that lives in the soil. They eat the roots of plants, particularly grass, creating bare patches in our lawns and devastating our vegetable gardens. Cutworms prefer to travel through moist soil, and the moths need bare patches of soil to lay their eggs. Generally, they will only predate or cause problems to the roots of poorly established or young plants. So, nurturing your plants for a little longer before planting them outside can help to make them more robust against cutworms.

An effective treatment for cutworms can be an application of nematodes. These little worms are available to purchase, and you can

A note on nematodes

Nematodes are microscopic worms that exist in healthy soils. Some of them are good news, some not-so-good news. As an agent for controlling pests, they can be excellent. However, they won't last long in poor soils. And healthy soils will already contain nematodes. Adding anything to your soils is problematic, as it creates an imbalance in the ecosystem, but to the best of my knowledge, at the time of this writing, nematodes do not cause undue harm, and they are certainly a better option than chemical treatments. However, if you can avoid using them, do.

get nematodes specifically for controlling cutworms. The nematodes enter a cutworm's body and deposit bacteria that kill the cutworm. Not a very nice way to die, I think we'll all agree, but cutworms fail to generate too much sympathy from gardeners.

The good thing about using nematodes on the cutworms is that they then ensure there are no bare patches in your garden for the moths to lay their eggs. You may be able to fix the problem with a single treatment.

MANAGING WEEDS

We all know them, those stubborn plants that grow persistently and consistently where we don't want them. There are three types of weed: annual, biennial and perennial. As with other plants, annual weeds will complete their lifecycle within one growing season; biennial weeds take two growing seasons; and perennial weeds will live for many years. Some weeds can be removed quite easily, and some are much more stubborn. Where you live will likely dictate what kind of weeds you have, and something that is considered a weed in one part of the world might be considered a fabulous ornamental in another. So, I won't be going into specific species. Instead we will cover a few rules of thumb when it comes to getting rid of weeds.

Annual weeds

Most of us can recognise annual weeds when we see them. They grow incredibly quickly and go to seed fast. Often, we can be in our gardens weeding and think that we've got everything and, when we turn round again, there they are, as though they've just grown in the few seconds when we looked away. To achieve this near miracle of growth, these plants like to feed off a specific type of nitrogen called *nitrate*. Soils contain two types of nitrogen: nitrate and ammonium. Most plants like a balance of nitrates and ammonium, preferring only a little nitrate. Fast-growing annual weeds, however, prefer mostly nitrates. In unhealthy soil, nitrifying bacteria are present in high numbers. These are bacteria that convert ammonium to nitrates. This connection between unhealthy soil and these bacteria means that, in disturbed and depleted soils, there is more nitrate. The high nitrate levels allow weeds to grow quickly and set seeds quickly. But as the soil becomes healthier, the populations of nitrifying bacteria go down, and so do the levels of nitrates.

MANAGING PESTS AND WEEDS

Therefore, one of the best defences against annual weeds is in fact, healthy soil. Healthy soil is covered in depth in Chapter 2, 'It all starts with soil' (page 31), and I encourage you to go back to that chapter. It will help you understand what healthy soil is, and how to achieve it.

Hoeing

Let me put in a humble vote for the good old-fashioned hoe. I highly recommend investing in a good-quality, sharp hoe. A Dutch hoe is my personal weapon of choice but to each their own. Having both a long-handled and a short-handled hand hoe is the best option. This allows you to quickly cut weeds off at ground level. Running a hoe around regularly is by far the easiest way of keeping on top of small annual weeds. If the surface of your soil is soft you will find it much easier, so employing the no-dig method and laying compost on the surface of your soil will give you a much easier surface to hoe.

Most annual weeds are pretty easy to get rid of. They're just really good at coming back. If you are persistent with getting rid of them, and specifically not letting them go to seed, then you will be able to keep them at bay. But it's good practice to employ other methods, such as keeping the soil covered with plants to prevent the proliferation of annual weeds.

The important thing to note about hoeing is that, if you are regular and persistent, you can seriously damage a weed's ability to produce energy. Weeds that live underground and put their leaves up all over your garden need to photosynthesise to generate energy for the plant. If you are routinely hoeing the leaves out before they have a chance to feed the plant, the health of the plant will be seriously impacted. Don't forget that before you see the leaves they will be just beneath the surface of the soil. If you have a

Hoeing is a quick and efficient way to deal with weeds.

fine mulch on top of your soil that you can run a hoe through, you stand a good chance of cutting the leaves off before they even reach the surface.

Weed suppressants

Another method for preventing annual, biennial and perennial weeds is using a weed suppressant such as cardboard or landscaping fabric. The material is laid over the weeds and acts as a physical barrier, preventing sunlight from reaching your weeds. After a few weeks with no sunlight, most weeds will start to grow pale and weak and eventually die. If you intend to remove the suppressant, as you would with fabric, you should first check that it killed off the majority of the weeds. Cardboard slowly breaks down in the soil, so you can cover it in compost and leave it. I recommend wetting it thoroughly so that if you need to plant through it you can easily poke a hole in it.

Typically, anything that is thin and feels like a light fabric won't do much against weeds. Many of us have seen dandelions brazenly pushing up through tarmac paths, so what's a bit of fabric? Thin suppressant can be torn easily, and the weeds push straight through. Alternatively, you may find that the suppressant does its job but starts to break down in situ, with weed roots adhering to its underside. Then it becomes even more of a problem when you need to replace it. In my garden I had to remove large amounts of this in long thin ribbons. It was doing nothing to suppress the weeds but was adding plastic to the garden and was an awful pain to get rid of.

Heavy-duty weed or landscaping fabric is slightly different; it is woven and quite heavy. It typically does a good job of suppressing weeds and is quite robust. This heavy-duty fabric should be laid on top of the soil, or over the weeds. However, it is a lot of plastic. I got some second hand from someone local and used that. That way I don't feel guilty about the plastic, and I can still use the suppressant. This kind of fabric works perfectly in the short term. It takes between six weeks and three months to do the job effectively, depending on the weeds you have, and simply won't work at all on some pernicious weeds. This heavy-duty fabric is useful, though, for suppressing weeds on a portion of the garden whilst you are working on another portion. It will prevent annual weeds from going to seed, and it will be easier to clear the weeds once you are ready to start work on that part.

MANAGING PESTS AND WEEDS

Cardboard is another effective weed suppressant. I like that cardboard is free and it isn't made of plastic. It also rots down, so it is generally used as a suppressant underneath some compost, then, after around three months, there is no sign of any cardboard left and the weeds disappear with it. You will need several layers of cardboard for a heavy-duty suppressant and, again, it won't work on all weeds.

Power of mulching

In simple terms, mulching helps create a barrier, blocking light to seeds, which can suppress some weeds. A deeper mulch will do a more thorough job where this is concerned. But nothing will stop some weeds!

However, a good-quality mulch will also improve your soil condition and bolster the soil ecosystem. A healthy soil ecosystem will regulate the amount of nitrates in your soil, which will, in turn, regulate weed growth. In unhealthy soil, nitrate concentrations will be high, but in well-balanced soil there will be both nitrates and ammonium that produce strong, healthy growth and don't allow weeds to take over.

A good mulch also provides us with a soft surface for our beds. This softness is great for hoeing as it reduces resistance, meaning that we can easily stay on top of the weeds. Just like running a vacuum cleaner over a floor every few days, we should run a hoe over a bed every few days, too, particularly if it is prone to getting weeds. This maintenance usually gets to the weeds before they even have a chance to reach the surface and start photosynthesising.

Perennial weeds

Perennial weeds come back year on year. They will often die back over winter and come back with more vigour and enthusiasm the following year. These can be a pain to deal with, but there are a few simple guidelines to get you started.

Weeds with long tap roots should be dug out. You can do this using either a garden fork, a weeding tool or a trowel. Ensure you get as much of the root out as possible because many weeds can regenerate from just a 1cm (½in) section of their roots. If you are not confident that your compost gets hot enough to kill them or if the weeds contain seeds, you can 'drown' them in a bucket of water for a week or so before adding them to your compost bin. Woody weeds will certainly need digging out or they will continue to regenerate. Get as much of the roots out as possible and remove the woody base and stem. They can usually be added to the compost without any worries.

Weeds that pop up everywhere should be regularly hoed. These are the kinds of weed that exist as one large plant underground and pop up leaves everywhere you look. They can be extremely persistent, so regular hoeing is essential. If you have applied a compost mulch to your garden, hoeing will be easier. If they are coming up through a lawn, instead you may want to regularly either mow or pull them.

Grassy weeds should be pulled up or suppressed using a weed suppressant. If you are pulling them up, remember that they grow from the area where the root meets the leaves, so make sure that you are removing that junction part of the plant.

Weeds with a deep tap root like this dandelion should have the entire root removed to stop any resprouting.

A note on herbicides

Herbicides create imbalances in ecosystems, which, if they were in balance, would naturally ward off weeds for you. Herbicides may be a tempting quick fix, but they are a long-term disaster for your garden and yourself. Some are known to increase your risk of cancer by more than 40 per cent if you're exposed to them. So, I encourage you to look instead for the visceral satisfaction of yanking a weed out and managing to get the whole root, simply letting some weeds exist, or of gently hoeing between rows of lettuce.

CHAPTER 9

Common problems and how to deal with them

Many of us will experience the same common problems with our new gardens. Here are a few of the problems that people ask about most often during my talks and on my social media channels.

Problem number 1: My garden floods during winter and goes solid during summer

This is common amongst those of us who live on heavy clay soil, myself included. Clay is made up of very tiny plate-shaped particles that overlap with one another, creating a dense soil. Water doesn't easily find its way through, making it slippery when wet. And when it dries out in the summer, it contracts, leaving areas of hard ground fissured with cracks. For the soil at my allotment, it only takes a few days without rain for this to happen.

Organic matter helps to glue particles in the soil together (once it has passed through some microorganisms and become sticky). It also feeds the life in the soil that moves through and breaks it up, creating spaces between particles and allowing water to infiltrate better. You should mulch with organic matter, preferably composted material because it inoculates the soil with more life and doesn't create habitat for the kinds of creatures we don't want in our gardens. Mulching serves the additional purpose of being very porous. It absorbs water well and helps to prevent evaporation from the clay beneath, which regulates water temperatures. Try it on a patch of your garden; even after some really hot, dry days, the soil underneath the compost will be softer and damper than soil left exposed. You may need to mulch once

GROW A NEW GARDEN

Before replacing plants that have been eaten, see if you can find where your pests are hiding and reduce habitat around or access to your plants.

a year for several years to fix the problem, but you will see gradual improvements over that time.

Areas that are particularly prone to flooding may be slightly lower down than other parts of the garden – even if you are unable to detect this by eye. Laying some topsoil down to raise the level and then compost on top of that should sort out problematic low patches over time. Gypsum can be used sparingly as a last resort to help clay soils to drain. Aerating the soil gently with a fork may help to introduce some air if your soil has been waterlogged for an extended period.

For more information on what to add to the soil, see the 'Soil amendments' section in Chapter 2 (page 57).

Problem number 2:
I don't have the money to get all this compost

Don't panic! Compost is free; you needn't pay for it. Of course, this is harder if you want it immediately because generally compost takes around 6 to 12 months to make at home. There are plenty of sources of compost out there, though.

Firstly, try asking at your local gardening club as they will be sure to know all the local dealers. They will likely be able to point you in the direction of a farm to collect some well-rotted manure or a source of green waste (compost made from discarded food, garden waste and so on).

Then, if you are unable to find either of these options, you can try hot composting to make your own compost quickly. As soon as you decide to set up your garden, do so with a mind to making plenty of compost. Save cardboard boxes, woodchip, weeds and food scraps to add to your compost, and you should end up with a system that provides you with enough compost to keep the garden going in future years. Don't forget; if the soil in your garden is OK and everything grows pretty well, there is no need to add additional organic matter. Simply having plants growing in your garden will help bolster your soil's health. Just ensure you keep soil disturbance to a minimum.

For more information on creating compost, see the 'Making your own compost' section in Chapter 7 (page 165).

Problem number 3:
Everything I plant in my garden gets eaten

This can be a seriously frustrating problem and, of course, the solution depends somewhat on what is eating your plants. Sometimes, it isn't very clear; other times, the little blighters leave slimy trails everywhere. Or have the audacity to eat your plants in broad daylight right in front of you. (I'm looking at you, rabbits!) Once you have determined the pest you're up against, you can put up appropriate protection for your plants. If the pest might fall under the category of 'creepy crawly' then the best defence is simply planting out larger plants. You can also try growing less palatable plants. As a rule of thumb, plants with hard, glossy, spikey or thick leaves will be less edible than ones with soft green leaves. Red leaves are often less tasty than green, too. Imagine yourself as a pest; if it looks tasty, it's more likely to get eaten.

You can offer physical protection to plants that are being eaten by larger creatures. For example, growing something small on a shelf will likely stop it from being eaten (too much) by mice. Cloches, nets, cages, fences and tunnels will not disrupt the ecosystem too much but may help to keep out larger critters that are becoming problematic to your plants. If you want to be nice about it, you can cage some of the plants and leave others to be enjoyed by the local wildlife; these are called sacrificial plants. You shouldn't need to continue doing this forever, and not all plants will need it, so take a common sense, reactive approach if you choose to do this.

For more information about figuring out your pest and how to protect against them, see Chapter 8, 'Managing pests and weeds', on page 185.

Problem number 4: My garden is on a slope, so any organic matter I put on it just slides down when it rains

This is where terracing can be very useful. If your garden is on a steep slope, it might be necessary to create some terraces or steps, so that you have some flat growing space. If the soil is good already (does it have a nice crumbly texture and a chocolate-brown colour?) you might not need to add any organic matter to the soil and you can get away with planting straight into the slope. This is particularly useful when your budget is tight. Don't forget that you can use some robust plants, such as trees and shrubs, to bind the soil in particularly steep areas; you may want to plan your garden with this in mind. If the soil isn't very rich in organic matter, you could try sowing some green manure seeds. These are short-lived plants that will help improve the health of the soil. Sow these when no rain is forecast, and water little and often to prevent them washing down the slope.

You can also grow creeping plants such as ivy. Plant them at the top or bottom of the slope, and they will take root as they climb up or trail down, adding organic matter and binding the soil.

COMMON PROBLEMS

Problem number 5: I have hardly any topsoil and can barely get my spade in

This is particularly frustrating but not insurmountable. To overcome this problem, you will generally need to use raised beds. These raise the level of your soil, giving your plants more space for their roots. The other option you have is to choose plants, such as alpines, that will thrive in these sorts of limiting conditions, particularly if it is prone to drying out. You will need to get them established with a good watering schedule at first, but you can turn areas of thin topsoil into a feature if you wish. Using both of these methods could create an interesting, attractive and low-maintenance garden.

Problem number 6: My soil is completely dead, there aren't any earthworms, it's useless

I guarantee that your soil is not dead, unless you have experienced a serious poisoning event, which is probably beyond the remit of this book. Most people who think their soil is completely dead are saying so because it is lacking in earthworms. If earthworms don't have anything to eat, you're right, they won't be coming to your soil. But once you start to add organic matter to your soil, they'll have something to eat, and they will start turning up and making baby worms. If you can find organic horse manure, this is a really good source of earthworms, and they will help inoculate your soil to speed the process up. Organic compost will also help to add microorganisms. (How do you think it became compost in the first place?)

For more information on taking a fresh look at your soil, see 'Assess your soil' in Chapter 2, page 46.

Problem number 7: My garden gets hot and everything bakes in summer

This is a common problem with south-facing gardens. The answer is usually to create some shade and to mulch. Mulching will help to trap moisture in the soil – if you mulch with something organic such as compost, wood or bark chip, this will also help to improve the permeability of the soil, helping it to hold more water. Plants can cope with high temperatures as long as they have enough water. Creating shade can be done by tactically planting some trees or by adding structures such as pergolas. Choosing plants that are more suited to warm conditions may also help. Generally, plants with thick and waxy leaves are better suited to hotter places – but be sure to check that they will still be hardy in your area's winter unless you want to keep moving them indoors!

For more information on adding shade to your garden, see 'Creating shade' in Chapter 5, page 128.

Problem number 8: My garden is full of rubble

Gardens containing rubble are common with new-builds. If the builders are using topsoil before laying it to lawn, they will often take a shortcut and not bother clearing the rubble from building the house before laying the topsoil. This is of course an awful thing for the poor, unsuspecting homeowner. But for the builder, it saves time and disposal fees.

There are a few ways you can deal with this kind of situation. The first is to raise the level of the soil you intend to grow in by using raised beds or containers. You can also simply add as much organic matter as you can to the entire garden or the part you want to grow on, which will at some point become a good environment for plant growth. This process is documented by David Montgomery and Anne Biklé in their book *The Hidden Half of Nature*. Anne relentlessly added woodchips and any other organic matter she could find to her garden, which was on old glacial scree (similar to rubble but wild-built). They were surprised by how little time it took to make their garden a thriving paradise. For a deep dive into how this addition of organic matter works, I highly recommend checking out this book. This method simply bypasses the rubble and still allows your plants plenty of space for their roots.

The second method is much more labour-intensive: dig it out. It is possible to reclaim your soil by gradually digging out the rubble (the big pieces at least), but I probably wouldn't opt to take this approach unless the area is very small. Bear in mind that you will need to fill the spaces, too, or else suffer a garden full of craters!

Epilogue

One day I would like to stop creating new gardens. I dream of a garden that I can lovingly cultivate over many years. I want to put my roots deep into the earth and watch my garden thrive year after year. That time hasn't come for me yet, but I consider myself fortunate amongst my generation to have a garden at all. Creating a garden is not a task that one generally seeks to finish when you have caught the love of gardening.

As with life, gardening is a journey that we can choose to partake in mindfully and joyfully. If you would like to follow more of my journey, please visit my social media channels under 'Sow Much More'. There, you can get in touch with me or just follow along.

I hope that your garden brings you as much joy as mine does to me and my family.

Acknowledgements

Thank you to Kenny Wilding-Raybould for steering me, for always inspiring me and for never tiring of my garden-design questions. And, of course, for his contribution to arguably the most important part of this book. And to my wonderful partner Alex, for ever so patiently lending a hand both in the garden and behind the camera.

Glossary

If you're new to gardening, some of the vernacular can seem a little intimidating. So here is a glossary of simple gardening terms to get you started or to build your horticultural vocabulary. This glossary is by no means exhaustive. I have intentionally kept it short so as to be beginner-friendly and not too intimidating. If you can get to grips with this list, you should be able to understand at least seed packets, plant labels and your Great-Aunt Elspeth when she's talking about propagating herbaceous shrubs. I hope it will also help you use this book if you're an absolute beginner.

Annual. A plant that completes its entire lifecycle, from seed to flower, within one growing season. They are usually easy to grow from seed and easy to collect seeds from at the end of the season.

Anther. The part of the *stamen* that contains the pollen.

Biennial. A plant that completes its entire lifecycle, from seed to flower, in two years.

Broadcast sowing. A method of sowing that involves scattering seeds rather than planting them in a precise location. This is commonly used in large agriculture or when planting meadows.

Compost. A mixture of ingredients used to add nutrients and organic matter to the garden. It can be made from a wide number of things, including grass clippings, horse manure, woodchip and kitchen scraps. Making compost requires a good balance of these materials to assist in their biological breakdown.

Cotyledon. The first leaf or leaves to emerge from a seed. These leaves usually look different from the true leaves.

Cross-pollination. Where a flower receives pollen from another flower rather than from its own anthers.

Cultivars. A plant that has been selectively bred for certain characteristics. The cultivar is usually shown as follows: *Genus species* 'Cultivar'. For example, *Rudbeckia* 'Goldsturm'

Damp-off, damping-off. A fungal infection that occurs in *seedlings* when their potting *medium* is overwatered. It appears as a sudden wilting of the plant as the stems rot and break.

Deciduous. Trees or shrubs that lose their leaves during winter. It is a form of protection for the plant against the cold.

Dormant. When a plant shuts down for a period of cold. *Herbaceous perennials* die completely back to the ground and are dormant predominantly below the soil, whilst *deciduous* trees lose their leaves and go through dormancy with bare branches.

Drill. A linear, shallow groove made into the soil to plant seeds into. A drill is commonly used to plant carrots and parsnips, amongst other crops.

Evergreen. A plant that does not lose its leaves in winter.

Foliage. A plant's leaves, collectively.

Full shade. An area that does not receive any sun at any part of the day. Some plants are suited for these conditions; however, many will not grow well in full shade.

Full sun. An area that is in direct sunlight for most of the day.

Germination. A plant sprouting from seed (botanical term).

Harden-off, hardening-off. The practice of gradually acclimatising a plant to the cold and wind by bringing it outside during the day and taking it in again at night. It is supposed to help with *transplant shock*.

Hardy. Able to withstand a prolonged period of cold or frost.

Herbaceous. A plant or part of a plant that is green and not woody. Most perennials and almost all annuals and biennials are herbaceous.

Humus. The organic component of soil. It is made up of organic matter decomposed by *soil organisms*.

Hybrid. A plant that has genetic material from two different varieties (gardening term). Hybrids are cross-pollinated and their seeds will often not produce plants that are the same as the parent. In other words, to get another plant like a hybrid the original parent plants must be cross-pollinated again.

Irrigation. The application of water to your plants. There are many methods of irrigation, including by hand using a watering can or

GLOSSARY

hose pipe, or automatically using an irrigation system like a drip system or a soaker hose.

Loam. In general, good soil. But specifically, loam is made up of roughly 20 per cent clay, 20 per cent silt, 20 per cent organic matter, and 40 per cent sand. This type of soil isn't very common but is highly prized. Note: do not try to amend your soil with sand to create loam; try to grow things that are appropriate for the soil that you have.

Medium. The substrate into which a plant is planted (horticulatural term). In most cases, the medium is soil or potting compost.

Mulch (noun or verb). Noun: material used for covering that soil. Verb: To cover the soil. (Common gardening term.) A mulch can be applied around just the base of a plant or over an entire garden.

Node. The point on a stem where a leaf, flower or branching stem attaches (horticultural term).

Nutrients. The key elements needed for plant growth. Usually, we use this word to refer to soil-based nutrients. The main nutrients (*macronutrients*) found in the soil needed for plant growth (and their chemical symbols) are nitrogen (N), phosphorus (P) and potassium (K). Other important nutrients (*micronutrients*) include calcium (Ca), magnesium (Mg), sulfur (S), boron (B) and zinc (Zn). Atmospheric nutrients (additional micronutrients) needed for plant growth are carbon (C), oxygen (O) and hydrogen (H).

Organic matter. Gardening term that can mean several things. Usually, it refers to anything that is plant-based, but increasingly it is being used to refer also to the life within the soil.

Partial shade. An area that is shaded for some part of the day.

Peds. The crumbly texture of soil. Aggregating soil particles form a ped.

Perennial. A plant that will die back in winter and grow back again in spring.

Rhizosphere. Where the plant roots interact with the biology in the soil to exchange carbohydrates and nutrients. Essentially, the plant's external gut.

Root ball. The main ball of roots at the base of a plant.

Rosette. A structure where leaves or petals are arranged in a circular manner around a centre point.

Seedling. A young plant, raised from seed. Usually, they only have a few leaves.

Self-pollinating. A flower that can transfer pollen from its own anthers onto its own stamen to reproduce. It does not need pollen from another flower to produce seeds. In other words, it can complete its reproductive cycle even if kept in isolation.

Shrub. A woody plant with several branches arising from near the ground, typically smaller than a tree.

Soil organisms. Creatures that live in the soil. These creatures can be anything from earthworms and ground beetles to bacteria, protozoa, nematodes and fungi.

Stamen. The male part of the flower, which ends in an *anther*. Flowers have multiple stamens.

Stigma. The female part of the flower that receives the pollen from the *anthers* (botanical term).

Systemic. Water-soluble and able to move through the tissues of plants. Regarding herbicides and pesticides, this attribute can make the whole plant toxic to bees and other beneficial insects.

Tender. When a plant is not able to cope with a frost. Some tender plants will not tolerate temperatures below about 5°C (41°F), and some will be fine until exposed to a prolonged period of sub-zero temperatures.

Tilth. A prepared soil surface. A fine tilth is made using a rake or hoe and a rough tilth by digging.

Transplant (noun or verb). To move a plant from one location to another. It usually refers to moving seedlings from a greenhouse to the garden. A plant can be called a transplant if it is grown in this way.

Tuber. A swollen root or underground stem that stores nutrients for the plant for the following season. A potato is a type of tuber. Dahlias and begonias also grow from tubers.

Waterlogged. Saturated with water. Most plant roots need oxygen for respiration, so waterlogged soil can kill some plants very quickly.

Index

A

acers, soil pH for, 41
acidic soil, 41
    DIY testing of, 44
    hydrangeas in, 41, 45*f*, 46
    lime added to, 42, 60
acidification, 42
aeration of soil, 149–52
    deep-rooted plants in, 151–52
    manual methods in, 141, 149, 206
    oxygen in, 156
    soil amendments in, 149–51
aerial photographs, 84, 87
aesthetics of raised beds, 108
aggregation of soil, 51, 53–57
    in clay soil, 51, 53, 59, 60, 61
    peds in, 55, 215
    in topsoil products, 58
alfalfa, 151
alkaline soil, 41, 42
    DIY testing of, 44
    hydrangeas in, 46
allotments, 8, 187
aminopyralid, 65, 165
ammonium, 200, 203
annual plants, 213
    seeds of, 158
    weeds as, 200–203
anther, 213, 216
aphids, 197–98
aquatic plants, 156
arbours, 123–24, 130
Armillaria mellea, 190

The Art of War (Tzu), 188
asbestos, 112
asters, 97
autumn months
    cuttings in, 175
    dividing plants in, 177, 178
    lawn care in, 141
    leaf collection in, 141, 159, 160
    observation of garden in, 86
awkward spaces, 133–35
azaleas, 41

B

bacteria, 34*f*, 54, 55
    in compost, 66*t*
    nitrifying, 200
bamboo, 125, 127
bare-root trees, 133
bark, in paths, 116
barrier management
    of pests, 198, 208
    of weeds, 115, 116, 202–3
basal cuttings, 172, 174
bedrock, 14*f*, 36
bees, 28, 30
benefits of gardening, 24–30
biennial plants, 213
    weeds as, 200, 202
Biklé, Anne, 210
biodiversity, 3, 25–30, 185
    in hedgerows, 122
    in meadow, 154
    trees increasing, 27, 28, 120
biogeography, 186

birch trees, 80, 96
    in existing garden, 27, 72, 121
birds, 30
    in biodiversity, 27–28, 30
    in food chain, 33
    in hedgerows, 122, 123
    as pest predators, 189, 198
    sounds of, 128
    and trees in garden, 27, 28, 120
black leaf spots, 181
blood meal, 38
blueberries, 41, 44
bokashi composter, 166
bone meal, 38
boom and bust pest populations, 193
brambles, 16
brassicas, pests of, 188–89, 192, 194–95
bricks, 114
Briza maxima, 128
broad beans, safety tests with, 63, 165
broadcast sowing, 213
Buddleia, 157
butterflies, 28, 198
    white cabbage, 188–89, 192, 194–95
butterfly bush, 157
buying plants, 94, 100, 179–83
    of annuals, 158
    colours in, 93–94
    invasives in, 101

buying plants (*continued*)
   of moss, 153
   for repetition, 94
   from reputable source, 181
   of trees, 122, 130, 133

C

cabbage caterpillars and butterflies, 188–89, 192, 194–95
Calibrachoas, 102*f*
canopy planting, 90
carbohydrates in soil, 35, 37, 40
carbon, 32, 51, 52*f*
cardboard as weed suppressant, 202, 203
carrots, 112
caterpillars, 198
   of cabbage white butterflies, 188–89, 192, 194–95
   in food chain, 33, 35
chamomile, 154
Chelsea Flower Show, 71*f*, 118
chemical management
   herbicides in, 63, 65, 204
   pesticides in, 187, 196
chicken pellets, 38
children
   connection to nature, 1, 2, 25, 28
   fencing for, 131
   and joy in garden, 30
   lawn for, 71, 151, 152
   planning garden for, 25, 28, 71, 73, 74, 88, 108, 131
   raised beds for, 108
Chinese meadow-rue, 103–4
chlorophyll, 42
chlorosis, 42
clay soil
   aggregation of, 51, 53, 59, 60, 61
   assessment of, 46–47, 51, 52*f*
   compaction of, 149–52
   drainage of, 53, 146, 149–52, 205–6
   gypsum added to, 61, 150
   improving structure of, 56–57, 205–6
   lawns on, 146, 149
   lime added to, 60
   mulch on, 205–6
   pH of, 42
   plant choices for, 93, 94
   sand added to, 59, 149–50
   in summer, 86
   in topsoil products, 58
clearing overgrown garden, 16
clematis, 125
climbing plants, 134
   cuttings of, 172, 175
   on fences, 97–99, 123, 125
   in hedgerows, 122
   for privacy, 122, 125
   transplanting of, 18
clothing, protective, 16
clover, 154
coastal areas, 52, 61, 75
coastal garden style, 76
colour
   of compost, 66*t*
   of compost ingredients, 166–67
   design scheme on, 92*f*, 93–94
   of hydrangeas, 41, 45*f*, 46
   personal preferences on, 74–75, 103–4
   relaxing, 118
   of roots in purchased plants, 182–83
   of soil, 47, 51, 52*f*, 208
compacted soil, 149–52
   drainage of, 141, 146, 149–52
   in lawns, 141, 142, 146, 149–52
   manual aeration of, 141, 149
   soil amendments for, 149–51
compost, 61–62, 66*t*
   for compacted soil, 150–51
   for containers, 61–62, 110
   cuttings in, 173
   definition of, 213
   fertiliser in, 66*t*
   fungi in, 65*f*
   green waste, 65–67, 112, 206
   homemade, 67, 160, 163, 165–71, 206
   on lawn, 140, 145, 150–51
   leaf mould, 141, 159, 160
   in no-dig gardening, 49, 50*f*, 201
   peat-free, 164–65
   purchased, 164–65
   for seeds, 141, 159, 160, 163, 164, 165
   testing safety of, 165
   weeds in, 167, 168, 169, 203
compost bins, 166, 168
composting, 165–71
   bay system, 169–71
   food scraps in, 166
   hot, 166, 168–69, 206
   leaves in, 141, 159, 160
   location for, 133
   mix of materials in, 166–67, 206
   temperature in, 166, 167, 169, 203
compost pile, 166, 167
compost sieve, 159, 163, 165
construction activities
   allotment sites in, 8
   and new-build gardens. *See* new-build gardens
   soil problems in, 13–15
containers, 109–10
   in awkward area, 134
   bamboo in, 125
   compost in, 61–62, 110
   for cuttings, 172, 173
   fertilisers for, 38, 110
   herbs in, 138
   for neglected garden plants, 18–19
   in not-your-style gardens, 19
   rootbound plants in, 182
   in rubble-filled garden, 210
   for seed starting, 158–59, 164
   soil in, 38, 61–62, 109, 110
   trees in, 110, 122, 133

INDEX

vegetables and fruits in, 137–38
watering of, 137–38
contaminants
  in leaves, 160
  in manure, 62, 63, 65
  in rubble, 112
contemporary style, 76
copper, 42
costs
  of allotments, 8
  of borders, 110
  of compost, 206
  of containers, 109–10
  in new garden, 12–13, 19, 72
  in raised beds, 109, 111–12
  of seeds, 160
  of trees, 122
  of walls, 132
Cotesia glomerata, 194–95
cottage gardens, 76
cottonwood, 127
cotyledons, 163, 213
courgettes, 138
COVID-19 pandemic, 194
cranefly larvae, 143–44, 188
creeping plants on slopes, 208
cross-pollination, 213, 214
cultivars, 214
curiosity, 1, 7
curves in design, 90, 91*f*
cuttings, 171–76, 181
  basal, 172, 174
  greenwood, 175
  hardwood, 172, 175–76
  nodal, 172, 173–74
  softwood, 172–74
  in water, 174
cutworms, 187–88, 191, 199–200
  in lawn, 143–44, 199
  predators and parasites of, 189, 191, 199–200

D

dahlias, 97, 101, 119, 190*f*
damping-off, 214
deciduous plants, 214
  fallen leaves of. *See* fallen leaves
  for privacy, 121, 123
  for shade, 130
deer, 137, 144, 198*f*, 199
design. *See* planning and design
digging soil, 54–57
  and no-dig gardening. *See* no-dig gardening
  for planting, 57
  recovery after, 55
dividing plants, 176–79
dogs. *See* pets
dormant plants, 214
drainage
  of clay soil, 53, 146, 149–52, 205–6
  French drains for, 148
  of lawns, 140, 141–42, 146–52
  plants improving, 151–52
  of raised beds, 108
  soakaways for, 147–48
  soil amendments improving, 149–51, 206
  soil compaction affecting, 141, 146, 149–52
  sources of problems in, 62, 205–6
drawing map of garden plan, 86–89
dream garden, imagination of, 16, 73–78
drill, 214
drones, 87
drought-resistant plants, 145
Dutch hoes, 201

E

earthworms, 34*f*, 35*f*, 49–50, 209
east-facing gardens, 81*f*, 82*t*
ecosystem, 3–6
  biodiversity in, 3, 25–30, 185
  health of, 185–86
  pest predators and parasites in, 5–6, 186, 189, 192, 194–95
  pests in, 4–6, 21, 185–86, 191–92
  in soil, 32, 33–36, 54–56
Entangled Life (Sheldrake), 40
epiphytes, 157
equipment and supplies
  for cuttings, 172, 174
  hoes, 201–2
  and material choices, 107–16
  for seed starting, 158–62
Erigeron, 97
Eryngium yuccifolium, 128
Euromic micro clover, 154
evergreen plants, 214
  for privacy, 121
existing garden
  infrastructure in, 79
  neglected, 16, 18–19
  not-your-style, 19–20
  overgrown, 15–18
  trees in, 27, 72, 87, 88, 121
exudates, root, 35, 37, 38

F

fallen leaves
  in compost, 141, 159, 160
  on lawn, 141
  on paths, 116
  on ponds, 135
  from privacy plants, 121
fences and walls
  climbing plants on, 97–99, 123, 125
  height of, 123, 125, 130
  for privacy, 120, 123, 125
  shade from, 130
  in wind, 131–32
fertilisers, 38–39
  in compost, 66*t*
  for containers, 38, 110
  for lawns, 139
  liquid, 37–38
  pollution from, 39
  in polytunnels, 49
fish meal, 38
fleabane, 96*f*
flooding problems, 146–48, 205–6

flowering plants
   annual, 158
   attracting insects, 100–101, 102f
   colours preferred, 74–75, 103–4
   cuttings of, 171, 173
   division of, 177–78
   and kale, 138
   nectar of, 101, 103
   in no-dig garden, 57
   pH affecting, 41, 45f, 46
   purchase of, 183
foliage, 214. *See also* leaves
food chains, 33, 35, 194
food production, 135–38
   childhood memories of, 119–20
   fruit trees in, 76, 78, 96, 137f, 138
   planning for, 12, 76, 135–38
   vegetables in. *See* vegetable gardening
food web, 34f, 35, 36–37
foot traffic on lawns, 142, 149, 150
formal gardens, 75f, 76
foxes, 186
French drains, 148
front-line plants, 190–91
frost date, last, 164
fruits, 30, 76, 78, 96, 137–38
full shade areas, 214
full sun areas, 214
fungi, 34f, 37f
   in compost, 40f, 65f
   diseases from, 181, 190, 214
   glomalin from, 54
   in healthy soil, 40f, 65f
   hyphae of, 54–56
   mycorrhizal, 39–41
   and soil aggregation, 54–56

G
garden centres, local, 181
garden journals, 183–84
gazebos, 128, 130
germination, 214
   light in, 163, 164
   temperature in, 160, 161, 163
   testing of, 161
   tips for, 162–64
   water in, 162, 163
glomalin, 54
Google Maps, 84, 87
grasses, ornamental, 96, 97, 118, 126f
   for privacy, 127
grass lawns, 139–54
   alternatives to, 145, 152–54
gravel paths, 116
greenhouses
   new plants in, 94
   seedlings in, 161–62
   vegetables in, 136f, 137
green waste compost, 65–67, 112, 206
greenwood cuttings, 175
gypsum, 61, 150

H
habitat
   fragmentation of, 186
   of pests, 188
hanging plants, 134
hardening-off, 214
hard landscaping, 88, 112–16
   in coastal style, 76
   existing, 72
   fences and walls in. *See* fences and walls
hardwood cuttings, 172, 175–76
hardy plants, 214
health benefits of gardening, 25
healthy plants, tips for buying, 179–83
healthy soil, 52–67
hedgehogs, 27
hedgerows, 122–23, 130, 131, 132
herbaceous plants, 214
herbicides, 63, 65, 204
herbs, 138
The Hidden Half of Nature (Montgomery & Biklé), 210
hoeing in weed control, 201–2, 203, 204
homemade compost, 67, 160, 163, 165–71
honesty (*Lunaria*), 128
honey fungus, 190
hot composting, 166, 168–69, 206
humus, 14f, 214
hybrid plants, 214
hydrangeas, 41, 45f, 46
hyphae, fungal, 54–56

I
identification of plants in new gardens, 18–19
Impatiens, 119
infiltration test of soil, 51
infrastructure of garden, 107
   existing, 79
   fences and walls in. *See* fences and walls
   for privacy, 123–24
insects
   in biodiversity, 27
   flowers attracting, 100–101, 102f
   in hedgerows, 122
   in meadows, 154
   in new-build gardens, 22–23
   in raised beds, 108
invasive plants, 101, 127
Irish moss, 153
iron, 42
irrigation, 145, 214–15. *See also* watering
ivy, 125, 208

J
Japanese knotweed, 58, 101
jasmine, 125
journals, gardening, 183–84
joy, 30, 165, 211

K
kale, 97
   in flower beds, 138
   pests of, 188–89, 192, 194–95

INDEX

L

landscaping fabric, 202
lawns, 139–54
    alternatives to, 145, 152–54
    benefits of, 152
    compacted soil in, 141, 142, 146, 149–52
    compost on, 140, 145, 150–51
    cutworms in, 143–44, 199
    drainage of, 140, 141–42, 146–52
    dry and hard, 144–45
    foot traffic on, 142, 149, 150
    in new-build gardens, 15, 146, 149
    patchy grass in, 140, 142–44
    reseeding of, 140, 141, 151
    seasonal care of, 139–42
leaf mould compost, 141, 159, 160
learning process
    on design, 69–70, 100
    journal on, 183–84
    on pests, 188–91
leatherjackets, 143–44, 188
leaves, 214
    chlorosis of, 42
    fallen. *See* fallen leaves
    of purchased plants, 181
    rustling sounds of, 127–28
lettuce, 191
lifecycle of pests, 188–89, 193
lighting, 119
lime, 42, 60
loam, 46–47, 215
local regulations on fence height, 123, 125
Lunaria annua, 128
Lysimachia, 96f, 97

M

macronutrients, 157, 158, 215
mammals, 27, 28, 198–99
manganese, 42
manure, 13, 62–65
    composting of, 206
    contaminants in, 62, 63, 65, 165
    earthworms in, 49–50, 209
    testing safety of, 63
mapping garden, 86–89
materials in garden, 107–16
    budget for, 12–13, 107, 109
    existing, 79
    in hard landscaping, 72, 88, 112–16
    for privacy, 123–24
    for raised beds, 90, 108, 109, 111–12, 113f
    uniformity in, 90
McWilliam, Gavin, 118
meadow mix, 154
mealy bugs, 183
Mediterranean gardens, 75, 138
medium, planting, 215
    for cuttings, 172, 173
    overwatered, 214
    of purchased plants, 182
    for seed starting, 141, 159, 160
mental health benefits of gardening, 25, 30
mice, 199, 208
micro-aggregation, 54
micro clover, 154
micronutrients, 157, 215–16
mindset, 23–24, 25, 30
Montgomery, David, 210
moss, 153
moth larvae in lawn, 143–44
mowing, 109, 139–40
mulch, 215
    on clay soil, 205–6
    and fertiliser use, 38
    in hot weather, 209
    in weed control, 203, 204
mycorrhizal fungi, 39–41

N

nasturtiums, 192
native plants, 103
natural appearance of garden, 95f, 97, 100, 127
nature
    biodiversity in, 27–30
    connection with, 1, 2, 28
    ecosystem in, 3–6
    in organic gardening, 2–3
nectar, 101, 103
neem oil, 196
neglected gardens, 16, 18–19
nematodes, 34f, 199
    as cutworm parasite, 189, 191, 199–200
netting, as pest barrier, 198
new-build gardens, 13–15
    insects in, 22–23
    lawns in, 15, 146, 149
    rubble in, 15, 210
    shade in, 128
new gardens, 11–30
    challenges in, 20–24
    definition of, 11
    mindset in approach to, 23–24
    neglected, 18–19
    of newly built homes, 13–15
    not previously gardened, 20
    not-your-style, 19
    overgrown, 15–18
    pests in, 4–6, 21–23
nitrates, 200, 203
nitrogen, 39, 157, 200, 203
node on stems, 215
    cuttings taken below, 172, 173–74
no-dig gardening, 54–56
    compost in, 49, 50f, 201
    flowers in, 57
    weeds in, 201
noise, reduction of, 118, 122, 127–28
north-facing gardens, 80–81, 82t
not-your-style gardens, 19–20
nurseries, local, 181
nutrients, 32, 36–39, 157–58
    definition of, 215
    in liquid products, 37–38
    in organic matter, 36, 158
    in rhizosphere, 39
    and soil organisms, 35, 36–38

O

observation of garden, 78–86
    seasonal differences in, 80, 83, 84–86
    of sunlight and shade, 78–84
organic gardening, 2–3
    ecosystem in, 4
    pests in, 185, 187–95
organic matter, 42, 56f
    for clay soil, 205–6
    for compacted soil, 149
    in compost, 49, 50f, 61, 166–67
    definition of, 215
    for lawns, 144, 145, 149
    nutrients in, 36, 158
    on slopes, 208
    soil content of, 46f, 47, 51
    on surface, 47–49, 50f, 57, 58–59, 205–6
    in topsoil products, 58
orientation of garden, 80–81, 82t
overgrown gardens, 15–18, 24
Owen, Jennifer, 27
oxygen, 156, 167

P

pampas grass, 127
parasites of pests, 5–6, 189, 191, 192, 193
    of cabbage caterpillars, 194–95
    of cutworms, 189, 191, 199–200
    wasps as, 189f, 194–95
parasols, 129–30
partial shade, 215
patchy grass in lawns, 140, 142–44
paths, 109, 114–16, 142
peat, 51, 52f, 164–65
peds, 55, 215
perennial plants, 215
    cuttings of, 172
    division of, 176–79
    dormant, 214
    transplanting of, 104

weeds as, 200, 202, 203–4
personal preferences, 74–75, 101–4
    in garden style, 19–20
    in plant choices, 74–75, 103–4
pesticides, 187, 196
pests, 4–6, 21–23, 185–200, 207–8
    in compost areas, 166
    in ecosystem, 4–6, 21, 185–86, 191–92
    in ecosystem problems, 191–92
    food preferences of, 189–90, 197, 208
    front-line plants in, 190–91
    habitat preferences of, 188, 196–97
    in lawns, 143–44
    learning about, 188–91
    lifecycle of, 188–89, 193
    in new-build gardens, 22–23
    organic management of, 185, 187–95
    parasites of. *See* parasites of pests
    predators of, 5–6, 186, 189, 192, 193
    on purchased plants, 183
    in raised beds, 108–9, 188, 196–97
    sacrificial plants in, 192–93, 208
    in vegetable gardens, 137
pets
    design for, 70, 71, 74
    fencing for, 131
    and lawns, 143, 151, 152
pheromones, 195
pH meters, 43
pH of soil, 41–46
    and hydrangea colours, 41, 45f, 46
    testing of, 43–46, 60
phosphorous, 39, 42, 157
photosynthesis, 42
planning and design, 69–105

of awkward areas, 133–35
of beds and borders, 110–11
changes to, 71–72, 104–5
for children, 25, 28, 71, 73, 74, 88, 108, 131
colour in, 74–75, 93–94, 97, 103–4
containers in, 109–10
curves in, 90, 91f
fences and walls in, 97–99
flowers in, 100–101
for food production, 12, 76, 135–38
goals in, 24
hard landscaping in, 88, 112–16
imagining dream garden in, 16, 73–78
learning process in, 69–70, 100
lighting in, 119
mapping in, 86–89
material choices in, 90, 107–16
mindset in, 24
for natural appearance, 95f, 97, 100, 127
for neglected garden, 18–19
for noise reduction, 118, 127–28
for not-your-style garden, 19
observation for, 78–86
for overgrown garden, 16–18
personal preferences in, 74–75, 101–4
for pets, 70, 71, 74, 131
plant choices in. *See* plant choices
for potential uses, 16, 24–25, 73–74
for privacy, 117, 119–27
of raised beds, 108–9
repetition in, 94, 97
for shade, 128–31
show gardens as inspiration in, 70, 71f, 118
sun exposure in, 78–84

INDEX

themes and styles in, 74–78
uniformity in, 90, 94, 97
water features in, 118
for wind sheltering, 118, 131–33
plant care, 155–84
    challenges in, 157
    compost in, 165–71
    journal on, 183–84
    of lawns, 139–54
    nutrients in, 157–58
    oxygen in, 156
    and propagation, 158–65, 171–79
    sunlight in, 155
    water in, 155–56
plant choices, 74–75, 92–101
    colour in, 74–75, 92*f*, 93–94, 97, 103–4
    for containers, 110
    for creating shade, 130
    for fences, 97–99
    flowers in, 100–101
    in hedgerows, 122–23
    height and spread in, 90, 94, 99, 110
    in hot dry conditions, 209–10
    invasiveness as concern in, 101
    for lawn alternatives, 145, 152–54
    native plants in, 103
    for noise reduction, 127–28
    personal preferences in, 74–75, 103–4
    plant communities in, 97, 100
    for pollinators, 100–101, 103
    for privacy, 96, 120–27
    for repetition, 94, 97
    for theme or style, 75–76
    in thin topsoil, 209
    for wildlife, 76, 103, 138, 154
    for wind reduction, 131, 132–33
pollination, 213, 214, 216
    attracting insects for, 100–101, 103
    parasitic wasps in, 194–95

polysaccharides, 54
polytunnels, 47, 49
ponds. *See* water features and ponds
Pope, Alexander, 69
poplar *(Populus)*, 127
pore spaces, 53, 55
potassium, 157
predators of pests, 5–6, 186, 189, 192, 193
pressure-treated wood, 113–14
privacy, 117, 119–27
    fences for, 120, 123
    plant choices for, 96, 120–27
propagation of plants
    cuttings in, 171–76, 181
    divisions in, 176–79
    seeds in, 158–65, 179
property documents, garden map based on, 87

Q

quaking grass, 128
Quammen, David, 186

R

rabbits, 137, 199, 207
rainfall
    acidification from, 42
    collection of, 145
    drainage of, 146–52
raised beds, 13, 107–9, 209
    costs in, 109, 111–12
    materials for, 90, 108, 109, 111–12, 113*f*
    pests in, 108–9, 188, 196–97
    in rubble-filled garden, 210
    soil in, 52, 111–12
    trees in, 111, 122
rats, 199
rattlesnake master, 128
resilience
    of ecosystem, 4
    of pests, 21
respiration, 36, 156
Reynoutria japonica, 58, 101

rhizosphere, 39–41, 215
rhododendrons, 41
rhubarb, 179
rocks. *See* stones and rocks
root ball, 215
rootbound plants, 182
root exudates, 35, 37, 38
rooting hormone, 172, 173, 176
root system, 36–41
    deep, 151–52
    in nutrient extraction, 36–39
    of perennial weeds, 203–4
    of purchased plants, 182–83
    rhizosphere of, 39–41, 215
roses, 101, 104, 125
rosette, 215
rubble, 15, 112, 210
Rudbeckia, 74, 103

S

sacrificial plants, 192–93, 208
salt in coastal areas, 61
Salvia, 74, 97, 101
sand
    as fill for raised beds, 112
    as soil amendment, 59–60, 149–50
    in topsoil products, 58
sandy soil, 46–47, 51
    aggregation of, 51, 53
    improving structure of, 56–57
    pH of, 42
    in summer, 86
    water in, 47, 53
sawdust, as fill for raised beds, 111
scale bugs, 183
Searle, Becky, 6–8
seasons
    lawn care in, 139–42
    observation of garden in, 80, 83, 84–86
seedlings, 215
    damping-off in, 214
    greenhouse for, 161–62
    temperature for, 161, 162

seeds, 158–65, 179
    broadcast sowing of, 213
    collected from plants, 160
    compost for, 141, 159, 160, 163, 164, 165
    drill in soil for, 214
    equipment for planting, 158–62
    germination of, 160, 161, 214
    for lawn areas, 140, 141, 151
    planting date for, 164
    planting depth of, 164
    purchase of, 160
    storage of, 160
    tips for starting, 162–64
    viability of, 161
    of weeds, 115, 116, 167, 168, 169, 200
self-pollination, 216
shade
    from arbours and gazebos, 123–24, 130
    creating, 128–31, 209–10
    for dry lawns, 145
    estimating areas of, 83–84
    full, 214
    garden map on, 88
    from hedgerows, 123, 130
    from house, 81
    orientation affecting, 80–81, 82t
    from parasols, 129–30
    partial, 215
    plant choices for, 93
    for south-facing gardens, 128, 209–10
    from trees, 71, 80, 83, 121, 130
sheds, 133, 134f
Sheldrake, Merlin, 40
shrubs, 214, 216
    cuttings of, 172, 175
    fruit, 138
    on slopes, 208
silvergrass, 127
slopes, 208
slugs, 2, 190f, 196–97

    eggs of, 67f
    in raised beds, 108, 188, 196–97
smart gardening, 2–3
snails, 196–97
    eggs of, 67f
    in raised beds, 108
soakaways, 147–48
softwood cuttings, 172–74
soil, 31–67
    aeration of. *See* aeration of soil
    aggregation of. *See* aggregation of soil
    assessment of, 46–52
    building health of, 52–67, 209
    carbon in, 32
    colour of, 47, 51, 52f, 208
    compacted. *See* compacted soil
    in containers, 38, 61–62, 109, 110
    dry and hard, 144–45
    ecosystem in, 32, 33–36, 54–56
    food web in, 34f, 35, 36
    importance of, 31–33
    infiltration test of, 51
    layers of, 14f
    loam, 215
    of new-build gardens, 13–15
    nutrients in, 32, 36–39, 157–58, 215
    peds in, 55, 215
    pH of, 41–46
    pore spaces in, 53, 55
    in raised beds, 52, 111–12
    rhizosphere in, 39–41, 215
    squeeze test of, 46–47, 51
    tilth of, 216
    water in, 32, 36, 156, 216
soil amendments, 57–67, 149–51
    compost, 61–62, 65–67, 150–51
    gypsum, 61, 150
    lime, 42, 60
    manure, 62–65

    sand, 59–60, 149–50
    topsoil, 57–59, 150
soil organisms, 216
    bacteria, 34f, 54, 55, 200
    in compacted soil, 149, 150
    in ecosystem, 33–36
    in food web, 36–37
    fungi, 39–41, 54–56
    in humus, 214
    in rhizosphere, 39–41
    and root exudates, 35, 37
    soil disturbance affecting, 54–56
    and soil health, 209
    and soil nutrients, 35, 36–38
    in topsoil products, 58
The Song of the Dodo (Quammen), 186
south-facing gardens, 80–81, 82t
    creating shade in, 128, 209–10
sowing seeds, 158–65. *See also* seeds
'Sow Much More,' 211
spring months
    cuttings in, 172
    dividing plants in, 177, 178
    last frost date in, 164
    lawn in, 139–40
    observation of garden in, 85–86
squirrels, 28
stamen, 216
stigma, 216
Stipa gigantea, 127
stones and rocks
    as fill for raised beds, 111–12
    in hard landscaping, 114
    in paths, 116
    in topsoil, 52
storage areas, 133
styles in garden design, 74–78
subsoil, 13, 14f, 15
substratum, 14f
sugars in soil, 35, 37, 40, 54
sulphur, 42
summer months

INDEX

cuttings in, 173
deciduous trees in, 121, 123, 130
hard ground in, 205
hot and dry, 209–10
lawn in, 140–41, 142–45
observation of garden in, 86
privacy in, 121, 123
sunlight and shade in, 80, 83, 84, 130, 209–10
transplanting in, 104
sunflowers, 151
sunlight, 155
   for container-grown plants, 109
   in fun sun areas, 214
   garden map on, 88, 89
   observation of, 78–81
   orientation affecting, 80–81, 82*t*
   plant choices for, 93
   in raised beds, 108
   seasonal changes in, 80, 83
   and shade areas, 83–84
   for vegetables, 135–37
suppliers
   of plants, 181
   of seeds, 160
systemic treatments, 216

T

temperature
   of compost, 166, 167, 169, 203
   for cuttings, 174
   for germination, 160, 161, 163
   plant tolerance of, 214, 216
   for seedlings, 161, 162
tender plants, 216
terracing of slopes, 208
terrain maps, 87
Thalictrum delavayi, 103–4
themes in garden design, 74–78
tilth, 216
topsoil, 14*f*
   depth of, 52, 57, 209
   in new-build gardens, 13, 15

as soil amendment, 57–59, 150
trailing plants, 134
transplanting, 216
   hardening-off in, 214
   from neglected gardens, 18–19
   in plan changes, 104–5
   shock in, 214
   of well-established plants, 105
trees
   bare-root, 133
   and biodiversity in garden, 27, 28, 120
   in containers, 110, 122, 133
   cost of, 122
   cuttings of, 175
   in existing garden, 27, 72, 87, 88, 121
   fallen leaves of, 116, 121, 135, 141
   by fences, 99
   fruit, 76, 78, 96, 137*f*, 138
   honey fungus of, 190
   in overgrown gardens, 16
   for privacy, 96, 120–22
   purchase of, 122, 130, 133
   in raised beds, 111, 122
   shade from, 71, 80, 83, 121, 130
   on slopes, 208
   staking of, in windy areas, 133
   transplanting of, 18
   as windbreaks, 132–33
trellises, 123, 125, 130
tropical gardens, 75
tubers, 216
Tzu, Sun, 188

U

urban areas, 5*f*, 89

V

vegetable gardening, 12, 76, 78, 135–38
   in allotments, 8
   in containers, 137–38
   daily sunlight for, 135, 137
   greenhouses for, 136*f*, 137

joy from, 30
no-dig, 57
pests in, 137
plan for, 88
purchase of plants for, 158
in raised beds, 111
verbena, 172
veronica, 97
vine weevils, 183
violas, 191
Virginia creeper, 125

W

walls. *See* fences and walls
wasps, parasitic, 189*f*, 194–95
water
   cuttings rooted in, 174
   and damping-off, 214
   drainage of. *See* drainage
   infiltration test of, 51
   plant need for, 155–56
   for seed germination, 162, 163
   and soil health, 32
   and soil pH, 42, 60
   and soil structure, 36, 51, 53
water features and ponds, 28, 29*f*, 118
   in awkward areas, 134–35
   biodiversity in, 28, 76
   in formal garden, 75*f*
   in neglected gardens, 18
   soothing sounds in, 128
   at West Green House, 68*f*, 69–70
watering
   of containers, 137–38
   of cuttings, 174
   and damping-off, 214
   of divided plants, 179
   of lawns, 140–41, 142, 145
   of seeds, 159
watering cans, 159
waterlogged soil, 36, 216
   in lawns, 140, 141–42, 146–52
water pollution from fertilisers, 39

water table, 146
weed killers, 204
  in manure, 63, 65
weeds, 200–204
  annual, 200–203
  barriers to, 115, 116, 202–3
  biennial, 200, 202
  in compost, 167, 168, 169, 203
  hoeing of, 201–2, 203, 204
  mulching of, 203, 204
  in neglected gardens, 19
  in not-a-garden gardens, 20
  in overgrown gardens, 16
  in paths, 115, 116
  perennial, 200, 202, 203–4
  purchased plants with, 183
  seeds of, 115, 116, 167, 168, 169, 200
  and soil health, 201
  in topsoil products, 58
west-facing gardens, 81*f*, 82*t*
West Green House fountain, 69–70
white butterflies, cabbage, 188–89, 192, 194–95
  parasitic wasp of, 194–95
white noise, 127–28

Wilding-Raybould, Kenny, 72–73
  on colour choices, 93, 97
  on curves in design, 90, 91*f*
  on learning process in garden design, 100
  on plant choices, 96–97, 99
  on plant communities, 97, 100
  on uniformity in design, 90, 97
wildlife, 76
  in biodiversity, 27–30, 76
  in ecosystem, 4, 21
  and fruit trees, 138
  in hedgerows, 122–23
  lawn alternatives for, 154
  in new-build gardens, 22–23
  as pests, 137, 207–8
  plant choices for, 76, 103, 138, 154
willows, 172
Wilson, Andrew, 118
wind, 118
  sheltering from, 131–33
wind chimes, 128
winter months
  cuttings in, 175

deciduous trees in, 121, 123, 130
flooding in, 205
lawn care in, 141–42
observation of garden in, 84–85
sunlight and shade in, 80, 83, 84, 121, 123, 130
transplanting in, 104
woodchips
  as fill for raised beds, 111
  in paths, 116
  in rubble-filled garden, 210
woodlice, 189
wood materials
  as fill in raised beds, 111
  for forming raised beds, 108
  in hard landscaping, 113–14
  in paths, 116
  pressure-treated, 113–14
wormer medications, 62

Y

yarrow, 97

Z

zinc, 42
zucchini, 138

About the Author

Alex Moorhouse

Becky Searle is an ecologist turned kitchen gardener. A self-taught designer, she has planned her own gardens from scratch. Her first garden as an adult was a tiny terrace outside a ground-floor flat. Since then, she has gardened in a 1960s house, a 1930s house and two new-build houses.

Becky has a monthly feature in *Kitchen Garden* magazine, for which she was shortlisted for 'Environmental Journalist of the Year' by the Garden Media Guild in both 2022 and 2023. She also writes for *Homes and Gardens* magazine, *Bloom* magazine and *Gardener's World*. Her own garden has been featured in the *Guardian* and *Modern Gardens* magazine.

Becky is @sow_much_more on Instagram.

the politics and practice of sustainable living
CHELSEA GREEN PUBLISHING

Chelsea Green Publishing sees books as tools for effecting cultural change and seeks to empower citizens to participate in reclaiming our global commons and become its impassioned stewards. If you enjoyed *Grow a New Garden*, please consider these other great books related to gardening and ecology.

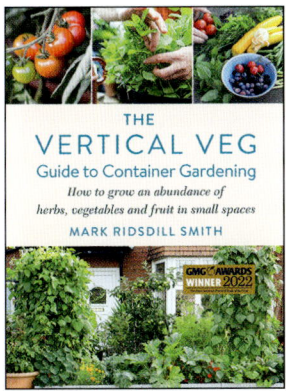

THE VERTICAL VEG GUIDE TO CONTAINER GARDENING
How to Grow an Abundance of Herbs, Vegetables and Fruit in Small Spaces
MARK RIDSDILL SMITH
9781645021506 (US)
9781915294609 (UK)
Paperback

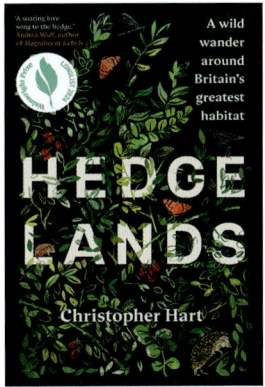

HEDGELANDS
A Wild Wander Around Britain's Greatest Habitat
CHRISTOPHER HART
9781915294470 (US)
Paperback
9781915294197 (UK)
Hardcover

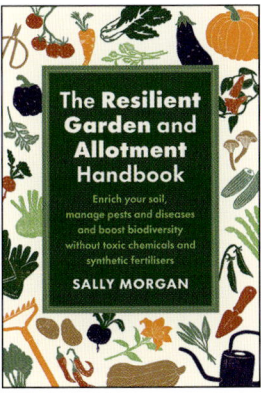

THE RESILIENT GARDEN AND ALLOTMENT HANDBOOK
Enrich Your Soil, Manage Pests and Diseases and Boost Biodiversity without Toxic Chemicals and Synthetic Fertilisers
SALLY MORGAN
9781915294562
Paperback

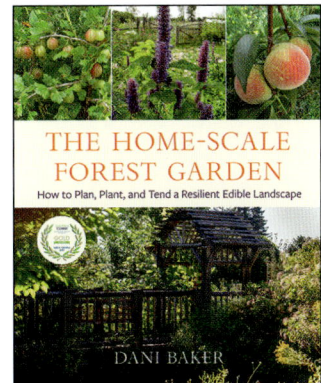

THE HOME-SCALE FOREST GARDEN
How to Plan, Plant, and Tend a Resilient Edible Landscape
DANI BAKER
9781645020981
Paperback

For more information,
visit **www.chelseagreen.com**.

the politics and practice of sustainable living
CHELSEA GREEN PUBLISHING

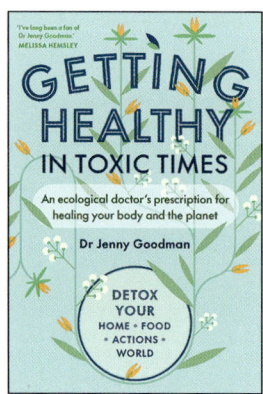

GETTING HEALTHY IN TOXIC TIMES
An Ecological Doctor's Prescription for Healing Your Body and the Planet
DR JENNY GOODMAN
9781915294333
Paperback

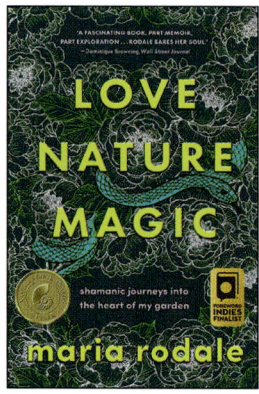

LOVE, NATURE, MAGIC
Shamanic Journeys into the Heart of My Garden
MARIA RODALE
9781645021681
Paperback

IN SEARCH OF THE PERFECT PEACH
Why Flavour Holds the Answer to Fixing Our Food System
FRANCO FUBINI
9781915294296
Hardcover

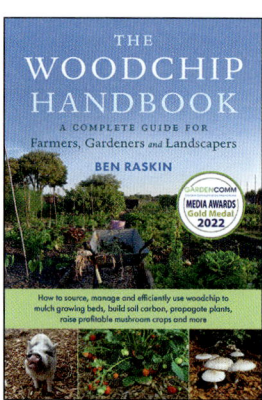

THE WOODCHIP HANDBOOK
A Complete Guide for Farmers, Gardeners and Landscapers
BEN RASKIN
9781645020486
Paperback

For more information, visit www.chelseagreen.com.